The Chipmakers

TIME
LIFE ®

Other Publications:
AMERICAN COUNTRY
VOYAGE THROUGH THE UNIVERSE
THE THIRD REICH
THE TIME-LIFE GARDENER'S GUIDE
MYSTERIES OF THE UNKNOWN
TIME FRAME
FIX IT YOURSELF
FITNESS, HEALTH & NUTRITION
SUCCESSFUL PARENTING
HEALTHY HOME COOKING
LIBRARY OF NATIONS
THE ENCHANTED WORLD
THE KODAK LIBRARY OF CREATIVE PHOTOGRAPHY
GREAT MEALS IN MINUTES
THE CIVIL WAR
PLANET EARTH
COLLECTOR'S LIBRARY OF THE CIVIL WAR
THE EPIC OF FLIGHT
THE GOOD COOK
WORLD WAR II
HOME REPAIR AND IMPROVEMENT
THE OLD WEST

This volume is one of a series that examines
various aspects of computer technology
and the role computers play in modern life.

UNDERSTANDING COMPUTERS

The Chipmakers

BY THE EDITORS OF TIME-LIFE BOOKS
TIME-LIFE BOOKS, ALEXANDRIA, VIRGINIA

Contents

17 ESSAY Into a Microworld
Electronic Powerhouses
ESSAY Creating Patterns of Logic

1

51 A Frontiering Industry
ESSAY Challenges of Fabrication

2

89 The Quest for Speed
ESSAY A Diversity of Chip Vehicles

3

120 Glossary
121 Bibliography
125 Acknowledgments
125 Picture Credits
126 Index

Into a Microworld

The invention of the microchip in 1958 changed the world. Without this electronic marvel—a small rectangle of silicon or similar material precisely etched and layered with circuitry visible only through a microscope—humankind's walks on the moon would, in all likelihood, still be aplanning. A wrist watch accurate to a few seconds a year would have remained beyond the means of all but the wealthy. Automobile engines, lacking chip-managed control of fuel mixture and ignition, would be a source of much more air pollution than computerized models are.

The tiny bundles of circuitry that accomplish such tasks come in marvelous variety. There are simple chips and complex ones—microprocessors that follow the instructions of computer programmers, specialized arrays that help manage data flow, memory chips that can be erased, and other memory chips whose contents are permanent. All chips lodge safe and secure inside hard plastic or ceramic shells that are easy to handle and to assemble into legions of digital devices. A prominent one is the motherboard of a personal computer, an assembly of chips that includes, among other things, the machine's microprocessor nerve center, some of the computer's memory, and channels of communication to the world of keyboards, monitors, and storage disks *(overleaf)*.

From a distance, the protective shells of the chips might be taken for buildings in an architect's model of a futuristic city or factory—an information-processing factory. Moving closer to "street level" reveals how the structures are attached to the circuit board. And at a magnification of 3,000 times, microscopic particulars of chip circuitry are discerned *(pages 14-15)*. Not at all the precisely defined, sharp-edged shapes implied by the precision required to manufacture components so small, they seem primitive in outline and construction, the result of processes that build a chip atom by atom, and molecule by molecule.

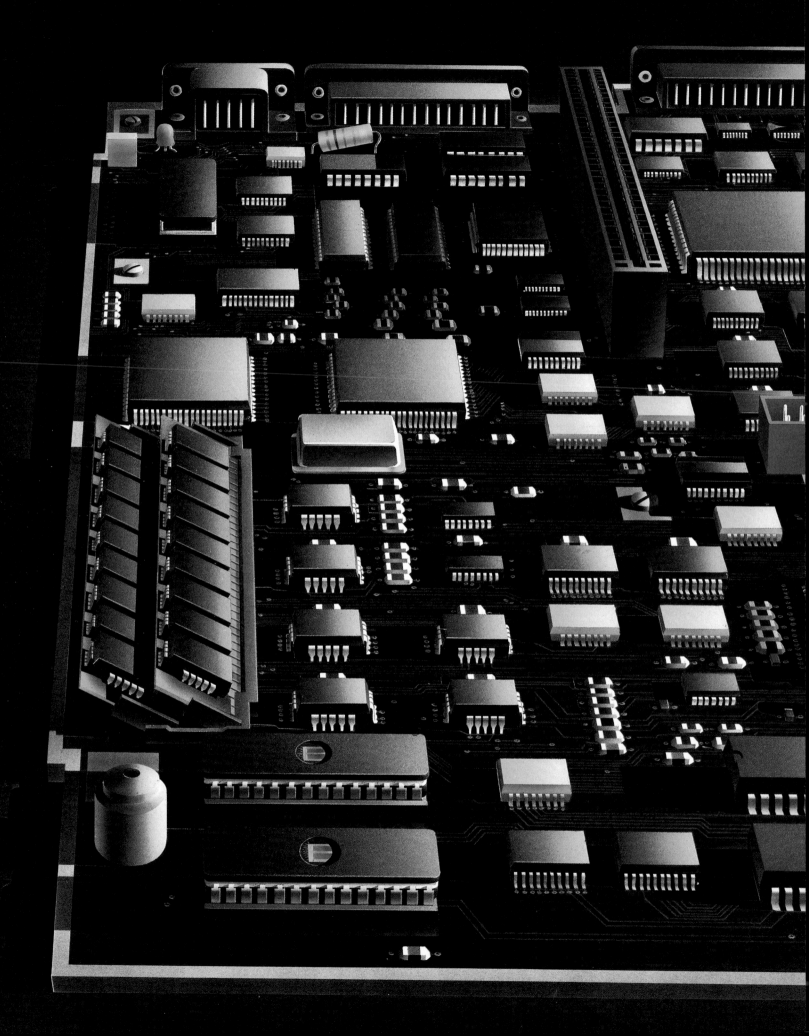

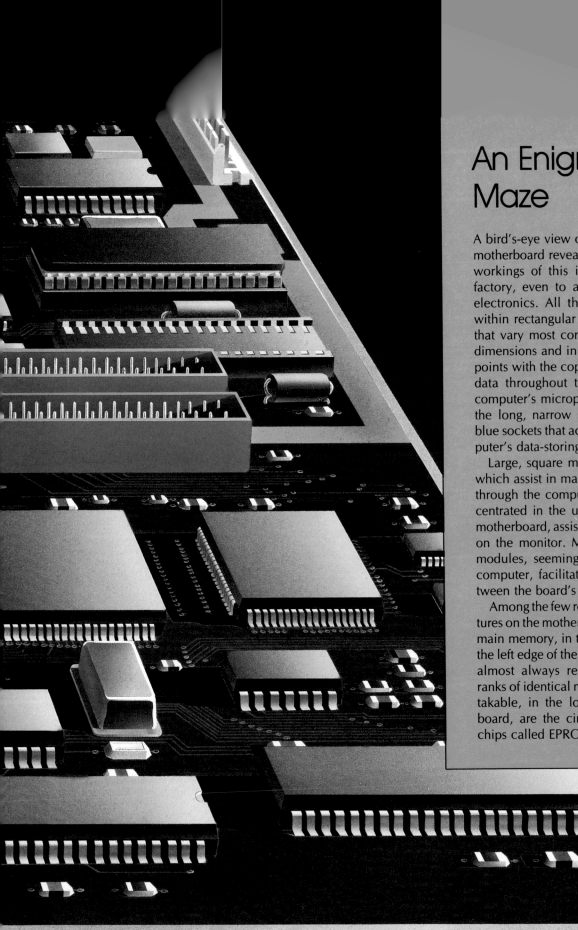

An Enigmatic Maze

A bird's-eye view of a personal-computer motherboard reveals little about the inner workings of this information-processing factory, even to an expert in computer electronics. All the chips are protected within rectangular slabs, called modules, that vary most conspicuously in external dimensions and in the number of contact points with the copper paths that conduct data throughout the circuit board. The computer's microprocessor resides inside the long, narrow slab behind the bright blue sockets that accept plugs for the computer's data-storing disk drives.

Large, square modules are gate arrays, which assist in managing the flow of data through the computer. Other chips, concentrated in the upper left corner of the motherboard, assist in displaying an image on the monitor. Many of the small chip modules, seemingly scattered about the computer, facilitate communications between the board's main functions.

Among the few readily identifiable structures on the motherboard is the computer's main memory, in this case installed along the left edge of the board. Main memory is almost always recognizable as parallel ranks of identical modules. Equally unmistakable, in the lower left corner of the board, are the circular openings of two chips called EPROMs (overleaf).

The View at "Street Level"

EPROM (erasable programmable read-only memory) chips hold information fundamental to a computer's operation, such as the sequence of commands automatically executed whenever the machine is switched on. To easily equip a motherboard with the most recent version of these instructions, the chip-module connectors, called pins, plug into sockets that have been soldered to the board.

Most modules, however, are soldered directly to the circuit board so that connections between chips and wiring are as secure as possible. Memory-chip modules *(tilted chips, far left)*, for example, are attached to the surface of a small board that is in turn plugged into a socket. A few others are pressed, like the EPROM sockets, into holes in a board before soldering.

Some pins, unneeded by the chip inside a module, do not connect with circuit-board conductors. Others are attached to short strips of copper that appear to go nowhere, ending in a silvery ring of solder. This feature is the top of a structure called a via—a shaft, usually solder-filled, that leads to one of as many as eight layers of conductors sandwiched inside the circuit board between layers of fiberglass.

From the circuit board, the pins lead inside a module to the chip itself. The quartz window in an EPROM module, through which ultraviolet light may be shone to erase the memory, affords a rare look at the microcircuitry of a chip.

1

Inside
a Module

Wires fine as silk arch gracefully forward from the ends of the pins to square contact pads that line the periphery of the chip. The wires are fused by heat to the pins and contact pads *(pages 110-111),* which are connected through microscopic amplifier circuits, represented as a maze of right-angled pathways in this illustration, to the rows and columns of memory cells that cover the chip. Empty pads between the wires may be used for testing the chip, reprogramming it, or as spares in the event that a pad proves faulty.

This module, like all others, is tightly sealed against the entry of moisture and other contaminants that could corrode the delicate wires and interrupt the flow of electrical signals from chip to pins. As an additional safeguard, these links are made of gold, which resists corrosion from dampness that might enter the module despite precautions.

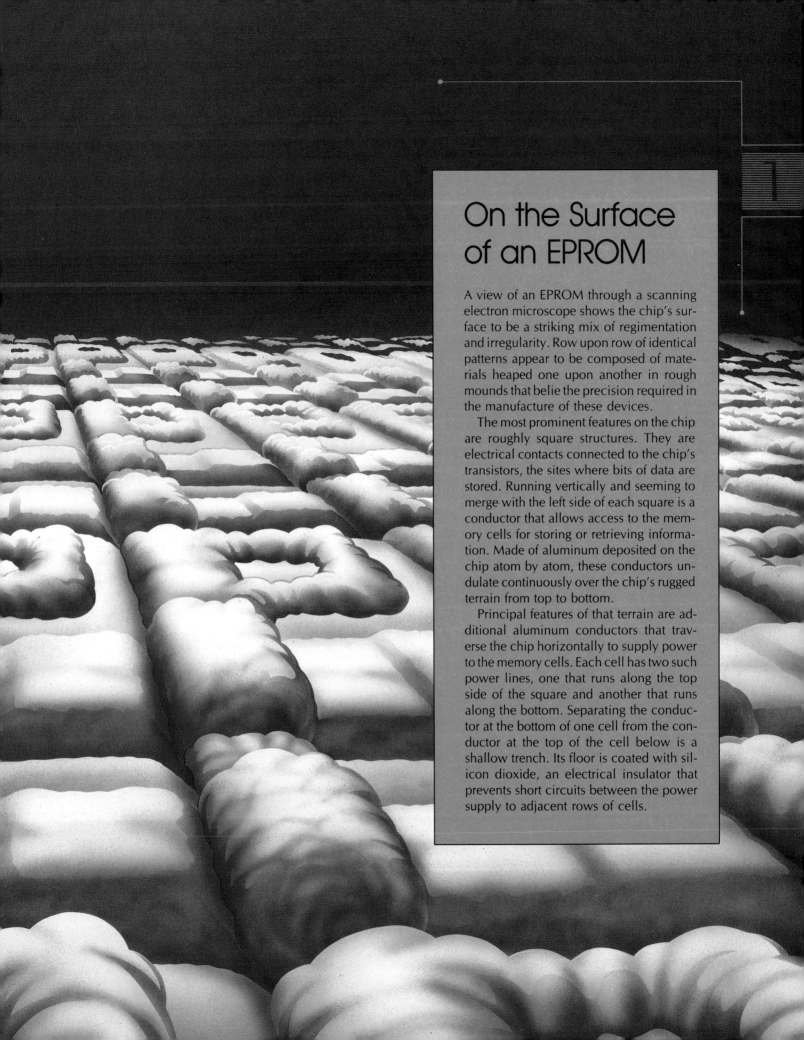

On the Surface of an EPROM

A view of an EPROM through a scanning electron microscope shows the chip's surface to be a striking mix of regimentation and irregularity. Row upon row of identical patterns appear to be composed of materials heaped one upon another in rough mounds that belie the precision required in the manufacture of these devices.

The most prominent features on the chip are roughly square structures. They are electrical contacts connected to the chip's transistors, the sites where bits of data are stored. Running vertically and seeming to merge with the left side of each square is a conductor that allows access to the memory cells for storing or retrieving information. Made of aluminum deposited on the chip atom by atom, these conductors undulate continuously over the chip's rugged terrain from top to bottom.

Principal features of that terrain are additional aluminum conductors that traverse the chip horizontally to supply power to the memory cells. Each cell has two such power lines, one that runs along the top side of the square and another that runs along the bottom. Separating the conductor at the bottom of one cell from the conductor at the top of the cell below is a shallow trench. Its floor is coated with silicon dioxide, an electrical insulator that prevents short circuits between the power supply to adjacent rows of cells.

Electronic Powerhouses

The invention of the transistor in 1947 by William Shockley, John Bardeen, and Walter Brattain, physicists at Bell Laboratories, launched a revolution in electronics, a field dominated since 1906 by the venerable vacuum tube. Invented by the American physicist Lee De Forest, the vacuum tube had reached its limits. A tube required large amounts of electricity to heat a filament inside it cherry red. One result was the release, from the filament, of electrons that other components inside the tube controlled in a number of ways. De Forest's original invention, a three-element device called a triode, could control the flow of electrons to a positively charged plate inside the tube. This ability made possible the first electronic digital computers which, like virtually every such computer since, relied on the zeros and ones of binary arithmetic to perform their feats of logic and mathematics. A zero could be represented by the absence of an electron current to the plate; a small but detectable current served as a one.

Other results of the hot-filament approach to electronics were less desirable. Tubes were bulky, and they generated so much heat that the cooling systems needed to prevent overheating in huge assemblies of the devices, such as were necessary to build even a modest computer, consumed nearly as much power as the tubes themselves. They were also unreliable; vacuum tubes tended to burn out like light bulbs. Indeed, the practicality of computers built with vacuum tubes was ultimately limited by tube-failure rates. A computer having thousands of tubes might operate for as short a time as one hour between malfunctions. If large computers were to become possible, a replacement for tubes was clearly essential.

The transistor promised an end to these problems—and it delivered. Within five years of the first working prototype, transistors were being made that could be expected to last indefinitely. They consumed much less power than vacuum tubes and therefore produced negligible amounts of heat. And they were so small that transistors were first used in hearing aids. Next, transistor radios were introduced. Soon, computer manufacturers replaced the thousands of vacuum tubes in their logic circuits with transistors. IBM introduced its first transistorized calculator in 1955. In place of 1,200 vacuum tubes normally required for a calculator of its capabilities, there were 2,200 transistors. Yet the new calculator was smaller, required no auxiliary cooling, and used 95 percent less power than the vacuum-tube model. Two years later, and ten years after the transistor's birth, the United States electronics industry produced almost 28 million of the devices worth $68 million.

The reliability and petite dimensions of transistors (the ones used as binary switches in IBM's computer fit inside protective canisters one-quarter inch or so in diameter and about one-quarter inch tall) led computer designers to contemplate machines of ever-increasing complexity, with more and more transistors. Additional miniaturization spurred these developments forward until evidence began to mount that further improvement in transistors would provide small returns for the effort invested.

The problem was that transistors, like other components of circuits, were individual items that had to be soldered together. As computers—and the circuits within them—became more complex, the number of connections skyrocketed, and so, too, did the chance of faulty wiring. Yet further miniaturization was already seen as an indispensable ingredient in the recipe for faster computers. Clearly, something had to be done about this serious shortcoming, or the computer's utility would be limited.

To the rescue came the integrated circuit, or chip. In essence, a chip is a collection of tiny transistors that are connected together as part of the process by which the transistor is created. The advantages are substantial. First, the need for soldering is much reduced. A chip containing 1,000 transistors, for example, may require as few as a dozen soldered connections to other components on a circuit board. Circuits containing a like number of discrete transistors would entail making at least 3,000 solder joints. Second, the reliability of the chip's internal wiring is comparable to the trustworthiness of the transistors themselves.

The evolution of the chip followed one rule from the device's invention through more than three decades of development: Pack ever-more-numerous electronic functions into increasingly tiny spaces, a process that was referred to by specialists in the field as increasing the scale of integration. The first integrated circuit consisted of just five electronic components and could perform only a very simple act—that of producing an alternating current when it was connected to a battery. By the late 1980s, a chip five millimeters square and half a millimeter thick could hold the electronics of a computer that fifteen years earlier filled a room five meters square.

In addition to saving space, miniaturization improves the performance of integrated circuits. As the dimensions of circuit elements and conductors shrank, so did the time required for electrical signals to travel through the circuits. Processing time decreased, and computers were able to perform more calculations in less time at lower cost. For example, in the early 1970s, the days of so-called large-scale integration (LSI), computers had a circuit density of about 5,800 circuits per cubic inch, while processing an instruction took less than a millionth of a second, and the tariff for handling a million instructions had dropped to two cents. By the 1980s, very large scale integration (VLSI) had come of age. Circuit density was greater than 600,000 circuits per cubic inch, while execution time was less than five nanoseconds (billionths of a second). The cost per million calculations had fallen to a hundredth of a cent. In addition, the average time between circuit failures in a typical computer had dropped from hundreds of hours in 1970 to years in the 1980s.

One way of describing this rate of progress is known as Moore's Law, named for Gordon Moore, one of the founders of Fairchild Semiconductor Corporation, the first chipmaking corporation. When asked to predict in 1964 how rapidly the chip would advance in the years ahead, Moore stuck tongue in cheek and predicted that integrated circuits would double in complexity every year. The pace of progress continues unabated, as manufacturers produce chips having more than a million components.

To achieve this degree of complexity, integrated circuits have undergone an astounding evolution on two fronts. On the science and engineering front, the development of new materials and new designs made possible—in theory at least—the shrinking of the transistor that led to tremendous aggregations of the devices. On the manufacturing front, methods and processes had to be invented and perfected for converting theory to practice. And chip evolution marches on. En route to an integrated circuit having a billion transistors, chip designers are pushing the limits of physics in an effort to build still-faster and more powerful integrated circuits, scientists are looking for and finding new and better insights into the special materials from which transistors are constructed, and manufacturers are inventing fabrication methods that can produce these feats of miniaturization reliably and economically.

IN THE BEGINNING

British scientist G. W. A. Dummer was the first to see the possibility of "wireless electrical circuits," as he called them. Dummer worked for his government's Telecommunications Research Establishment at Malvern, England, where he headed up the routine work of a testing group nicknamed the Dumpire in his honor. He meant by "wireless," not a radio (the common meaning of the word among Britons), but electrical circuits free of connecting wires and soldered joints. Addressing a group of engineers in Washington, D.C., in 1952, Dummer offered a good description of what a chip would be: "It seems now possible to envisage electronic equipment in a solid block with no connecting wires. The block may consist of layers of insulating, conducting, rectifying, and amplifying materials, the electrical functions being connected directly by cutting out areas of the various layers."

As revolutionary as Dummer's idea sounded at the time, it was actually a logical extension of the state to which the transistor had evolved. Transistors being manufactured in 1952 were of a type called junction transistors. Invented a few years earlier by William Shockley at Bell Laboratories, the device was built in layers, like a sandwich. Bread and filling alike were thin slices of germanium or silicon, nonmetallic crystalline substances and members of a small family of materials called semiconductors. Unlike insulators such as glass, which always resist the passage of electricity, or conductors such as copper, which always let it pass with little inhibition, semiconductors can be made to conduct electricity or not.

In pure form, semiconductors do not readily permit the passage of electricity. For that to occur, trace amounts of impurities, or dopants, must be combined with the material. Depending on the nature of the impurity, semiconductors can be made to carry current in the ordinary way—that is, by means of a flow of electrons as if through a wire—or by means of so-called electron holes, positively charged "absences of electrons" that move through the material much as electrons do.

For example, phosphorus, a commonly used dopant, has one more electron in its atomic structure than is necessary for an atom of the dopant to fit into the crystalline structure of, say, silicon. This electron is free to migrate through the material in response to voltage from a battery, just as free electrons in a metal wire do. Semiconductors in which there is a surplus of electrons—and a flow of

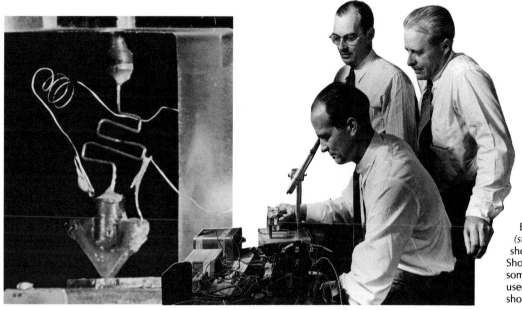

Bell Laboratories scientists John Bardeen and Walter Brattain *(standing, left to right)* look over the shoulder of colleague William Shockley, seated at a microscope and some of the other apparatus that was used to invent the first transistor, shown at far left about half life size.

negative current—are called n-type semiconductors. P-type semiconductors are those that have been doped with a chemical such as boron that has one too few electrons for a perfect fit in a silicon crystal. This deficit leaves a hole, a place in the crystal where an electron would fit if one were available. Connecting such a material to a battery causes neighboring electrons to fill these holes. That, however, creates a new, positively charged hole. In effect, then, a hole is a positive charge carrier that can carry current just as an electron can. In this type of material, there is no continuous movement of electrons, only holes. It is called a p-type semiconductor.

Shockley's junction transistor had a slice of p-type material formed between two slices of n-type material. The top slice, called the emitter, was smaller than the other two slices, so that the completed device had somewhat the appearance of a landscape in the southwestern United States—a low mesa rising from a flat plain. Electricity flowed from the emitter to the bottom slice, known as the collector, only when a positive control current was applied to the middle slice, or base. Scientists call this kind of transistor a bipolar type because part of this current was supplied by the transfer of electrons (the negative pole) and part of it by the holes (the positive pole).

Within limits, an increase in current through the base produced a larger increase in current at the collector. This quality was handy for computers. Low current at the base yielded low current at the collector, a zero in binary terms. A higher current at the base and a higher current at the collector signified a one. The transistor required three wires, one connected to the emitter, one to the base, and the third to the collector. The conductors were in turn soldered to other devices—perhaps other transistors—to make a circuit.

Bell Laboratories, rather than keeping the new developments to itself, made them available to virtually any company that expressed an interest. Surveying this technology, Dummer proposed that, instead of each mesa having a plain of its own, numerous small mesas should be built on a single, broader plain. As a part of the process, the transistors and other electronic devices constructed

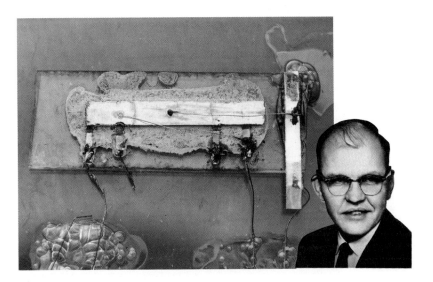

The first integrated circuit, invented in 1958 by Jack St. Clair Kilby of Texas Instruments, combined all the basic circuit elements in a single piece of germanium less than half an inch long *(horizontal bar)*. The device was held together by glue, with gold wires providing interconnections.

at the same time could be linked together to form useful electronic devices of all descriptions. Shortly after his speech in the U.S. capital, Dummer's hopes for making wireless transistors were dashed when his agency, because of shortages of silicon, limited the supply of that material to what they considered more critical projects. Though he continued his campaign, he never succeeded in getting his country's support, and he never built a working chip. A single model of an integrated circuit, constructed of silicon in 1957, was the culmination of Dummer's ten-year crusade to get someone to believe in his wireless circuits. "Dummer," a colleague said later, "was preaching the gospel of integrated circuitry long before anybody, including Dummer, had the slightest idea how you could actually do this." After his experience, Dummer decided to retire and write books. Had he been a better preacher, the chip might have been a reality sooner.

AWASH IN TRANSISTORS

By the time Dummer abandoned his project, the complexity of transistorized electronics had increased dramatically, and there was no end in sight. A tyranny of numbers threatened, in which the promise of transistors would not be fulfilled because of the overwhelming multiplicity of connections that were required to link huge quantities of transistors together. System designers, using transistors to replace vacuum tubes, had found that the reliability of transistors made possible the construction of increasingly complex computers—to the point where the reliability of soldered connections became the issue. That is, the probability of a loose connection somewhere in the maze of wires linking numerous transistors exceeded the likelihood that one of the devices would fail. Electronics designers, having tasted the benefits of increased system complexity, would not be satisfied until this new obstacle to further complexity had been surmounted.

Space researchers were also uneasy. Dreaming of the day when they might send a rocket to the moon, they envisioned a guidance computer that would contain 10 million components. Every component and every connection in those circuits would have to work when called upon, an extremely unlikely proposition with the transistor technology then available. Even if perfect operation could be guaranteed, the computer would be enormous, much too large to fit inside a rocket of practical size. The large size was objectionable for another reason. Electrical impulses being sent from point to point in such a computer, even at nearly the speed of light, encounter delays that cannot be tolerated in the control of a rocket, where the ability to solve complex control equations in milliseconds is essential.

The U.S. armed forces were particularly alarmed that the maze of soldered connections would become so unreliable as to be ill-suited for battlefield electronics, when weapons must function the first time, every time. Foreseeing difficulties as early as 1950, the navy had started Project Tinkertoy. The goal was to develop compact electronic modules that could be plugged together, like the popular wooden toy, to make a variety of electronic systems: radios, radars, fuses for bombs and artillery shells, and computers. Based on vacuum tubes rather than transistors, Tinkertoy was abandoned. In 1953, transistors became the basis for a similar program by the army, in which the assemblies were called micromodules. Transistors and other components were to be assembled onto tiny ceramic wafers about the diameter of a pencil. The wafers were to be stacked together, linked with wires around the edges. The stacks of wafers would then be plugged into circuit boards. Trial circuits actually worked, but the Micromodule Plan did not reduce the number of soldered connections, it merely changed their appearance.

The army had contracted with the electronics firm of Texas Instruments to press forward with the Micromodule Plan. One of the engineers working on the project was named Jack St. Clair Kilby, who had come to Texas Instruments after working for several years at Centralab in Milwaukee. That company had been involved in Project Tinkertoy, so Kilby knew of the earlier project's shortcomings firsthand and he felt that not much improvement had come with Micromodule. The army's approach was an example of what he described as getting the wrong answer by posing the wrong question. The Signal Corps had asked how these components could be connected more easily; Kilby asked how the connections might be eliminated.

Kilby was an inventor at heart, a man whose life revolved around developing practical solutions to sticky problems. He had yet to hear of the Englishman Dummer when, in July of 1958, while almost everyone else at Texas Instruments was away on vacation, Kilby started thinking about what he could do with a piece of silicon to overthrow the tyranny of numbers. Kilby, it seems, was much more familiar than Dummer with the basics of semiconductors. For example, from his work on micromodules, he knew that the semiconductive properties of doped silicon could be used to make transistors and even a second useful electrical device, called a diode, that provides for a one-way flow of current through a circuit. Kilby was also aware that researchers had recently discovered how to change silicon to silicon dioxide, or quartz—one of the best electrical insulators known—directly on a wafer of silicon. The process would allow both conducting and insulating areas to be created on the same slice of material to connect or isolate different components. Kilby also thought that he could make a silicon resistor, the electronic component used in circuits to limit the flow of current. He realized, too, that capacitors—electronic components used to store and delay current—could be formed at the junction between regions of n-type and p-type silicon.

Kilby had figured out how to make four of the five basic building blocks of all electronic circuits—transistors, diodes, resistors, and capacitors—all from silicon. (The fifth building block, the transformer and other inductive devices, is fundamentally incompatible with semiconductor technology; where an inductor is necessary it remains a separate component in the circuit.) "Nobody," Kilby said later, "would have made these components out of semiconductor material then. It didn't make very good resistors or capacitors, and semiconductor materials were considered incredibly expensive. To make a one-cent carbon resistor from good quality semiconductor seemed foolish."

Foolish or not, Kilby saw where these ideas led. If he could put all the elements of a circuit together on the same piece of silicon and connect them by wires laid on the surface of the chip he would have a monolithic device, his term for the "solid block" of electronic circuitry that Dummer had envisioned nearly a decade earlier. Kilby recorded the idea in his notebook in late July and had the first crude chip ready for testing on September 12, 1958. Less than half an inch long and narrower than a toothpick, the world's first integrated circuit contained only five components connected by wires sticking out of a sliver of germanium. (With most Texas Instruments employees on vacation, Kilby could not find a suitable piece of silicon.) But the device worked. Texas Instruments announced the chip's discovery in January 1959.

MULTIPLYING THE TRANSISTOR

As events unfolded at Texas Instruments, physicist Jean Hoerni, a founder of California's Fairchild Semiconductor, developed a new type of junction transistor that would soon displace the sandwich type that Kilby had used to make his chip. Hoerni found that by nesting layers of p- and n-type silicon only a few atoms thick, one layer inside the other like mixing bowls, he could make a transistor that was better than one having the broad area of contact achieved by stacking one kind of silicon on top of another, mesa fashion. He also made use of the discovery that small areas of a silicon crystal, doped in its entirety during manufacture to be either n-type or p-type, could later be redoped to become the opposite type.

Capitalizing on these ideas, the physicist was able to create a flat transistor that somewhat resembled a three-ring bull's-eye. The outer ring, a contact to the collector underneath, was made of the silicon on which the transistor was to be built, n-type for example. The middle ring, or base, consisted of a circle on the wafer that Hoerni redoped to become p-type silicon and then doped again to make a center spot of n-type silicon that served as the emitter. After doping the appropriate areas, Hoerni coated the transistor with a thin layer of silicon dioxide, leaving tiny holes in this protective layer that provided convenient contact points for electrical leads.

Hoerni's invention, called the planar transistor because the surface was flat, had several advantages over the mesa, junction transistor. With mesa technology, the raised surface of the chip provided niches that sometimes collected

The first planar (flat) transistor, shown with its inventor, Swiss physicist Jean Hoerni of Fairchild Semiconductor, was about 6/100 inch in diameter. Simpler to make than the devices that preceded it, the planar transistor became the basis for the first mass-produced chips *(opposite)*.

dust that could contaminate the chip, and the wire that was connected to the emitter atop the mesa often became dislodged. Most important, the planar transistor was inherently more reliable than a junction transistor and was more economical to produce.

Robert Noyce, a colleague of Hoerni's at Fairchild, saw great potential in the physicist's approach to making transistors. First, he realized that the technique might also work for making other electronic components on a silicon slab. Second, since Hoerni's transistor was flat, Noyce proposed to do away with the connecting wires by creating narrow conducting channels on the silicon's surface. Except where they were to be attached, they would be electrically isolated from other chip features by Hoerni's insulating layer of silicon dioxide.

Noyce's chip was the prototype for all that followed. It was elegant and it worked. More important, it could be mass-produced using techniques that were already accepted in the semiconductor industry. By 1961, both Fairchild Semiconductor and Texas Instruments were using the planar process to make chips, and both companies were marketing integrated circuits on a large scale. The era of the chip had arrived.

A COOL RECEPTION

At first, the electronics industry and the military greeted the integrated circuit with considerable skepticism. The navy and air force, for example, rejected Texas Instruments' original commercial circuit out of hand. The army gave the company a small development contract only in the belief that the chip could be assembled into micromodules.

Nonetheless, integrated circuits had much to offer. They were reliable. Chips shipped to customers almost always functioned as promised. Integrated circuits were also faster—and much more compact—than circuits made of individual transistors, capacitors, and the like. They also used less power, generated far less heat, and could withstand the effects of shock and vibration better than the soldered connections used to link discrete components.

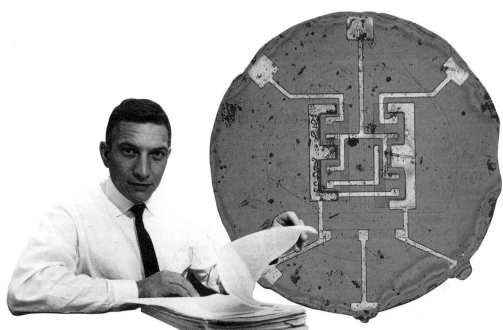

Early in 1959, Robert Noyce of Fairchild constructed an integrated circuit on a thin slice of silicon. Built-in metal conductors *(white)* linked to cone-shaped transistors and bar-shaped resistors *(blue)* eliminated the wires of Kilby's chip and made for easier mass production.

But there was still an overriding difficulty: The chips were too expensive. The first models cost more than circuits designed with individual components, even considering the higher assembly costs of the established technology. For example, Texas Instruments' first commercial circuit, a device called a flip-flop that could be used to store a single bit of data, cost $450. A similar circuit could be assembled from discrete components for about five dollars. The high price was probably justified. Designing chips was a new and time-consuming discipline; engineers had to learn how to combine components and wiring into an integrated whole—a concept that Kilby had named the monolithic device—rather than considering function and form as separate issues. Manufacturing the devices was no less demanding; imperfect processes led to many faulty chips. As few as 10 percent of the chips produced could be sold.

BOOSTED BY THE MOON SHOT

Too expensive to penetrate mainstream electronics markets such as telecommunications, the chip seemed stuck in the starting gate. Without an immense increase in demand, the volume of chips produced would remain small and the prices would stay high. But in the spring of 1961, the situation took a dramatic turn for the better. In May, President John F. Kennedy addressed a joint session of Congress. "I believe we should go to the moon," he said. "I believe that this nation should commit itself to achieving the goal, before the end of the decade is out, of landing a man on the moon and returning him safely to earth." The dream of the rocket scientists, to be named Project Apollo, was about to come true, and the solution to the 10-million-piece computer was to be the silicon chip.

"The space program badly needed the things that an integrated circuit could provide," said Kilby. "They needed it so badly they were willing to pay two times to three times the price of a standard circuit to get it." With price no object, suddenly there was a large market for chips. Almost overnight, the government signed contracts for new integrated circuits to be used in guiding the spaceship to the moon and back.

Thanks largely to NASA orders—and with a little help from the U.S. Air Force and the U.S. Navy, whose Minuteman II and Polaris intercontinental ballistic missiles incorporated integrated circuits beginning in 1962—chip production quadrupled in every year between 1963 and 1966. By the time the lunar lander *Eagle* touched down on the moon on July 20, 1969, the Apollo program alone had purchased more than one million chips. The integrated circuit was king of the electronics industry.

Predictably, new markets opened when chip prices plummeted—by nearly 50 percent annually—as a result of increased production. In 1963, the price of an integrated circuit averaged about $32.00. The next year, that figure was $18.50. A year later, the price fell to about $8.33.

With the money from the Apollo program and military contracts, Texas Instruments and Fairchild Semiconductor, as well as other companies that had entered the chipmaking business, were able to invest the large sums needed to improve design and fabrication techniques. Researchers at Motorola, for example, discovered a compelling reason to reduce the size of a chip: Smaller chips were more likely to work than bigger chips, because they are less likely than larger ones to run afoul of flaws in the silicon they are made of or to encounter dust particles. If ten chips are constructed on a wafer having five defects of dust, only five would work. But if forty chips could be fit onto the wafer, thirty-five of them would work—a sevenfold increase in output from a fourfold reduction in chip size.

IMPROVING INTEGRATION

In practice, the results were even better. Engineers discovered that production failures, caused by dust particles and other defects in the silicon, are not randomly distributed over the surface of the silicon wafers used to make chips. On the contrary, these faults are confined to limited areas of a chip. On a forty-chip wafer, the flaws are more likely to occur in an area covered by two chips instead of five. Thus, reducing the size of each integrated circuit had an even better payoff than expected. Of course, making smaller circuits took a considerable amount of effort, but driven by these production economies, improved fabrication techniques were developed rapidly.

One area of improvement was in the techniques used to build circuits onto silicon wafers. Essentially, the process is one of stenciling layer upon layer of conductors and insulators on top of the doped silicon wafer. Finer detail in the stencils, created by projecting patterns of conductors and insulators on light-sensitive stencil material, led to better definition of the circuit elements and allowed greater integration. To control the biggest gremlin in chipmaking—dust—clean rooms were developed that contained fewer than 100 dust particles per cubic foot. In comparison, the dust level in a modern hospital room is about 10,000 particles per cubic foot. Special clothing was designed for the plant workers to prevent dust, lint, dandruff, and flakes of skin from falling onto the wafers. New circulation systems were created to continuously filter and recycle the air in a facility *(pages 78-79)*.

As miniaturization progressed, the cost of each electronic component in a chip

Magnified 8,750 times by a scanning electron microscope, metal conducting lines—tinted silver and copper in this image and insulated from each other by a transparent layer—crisscross the surface of a memory chip. The metal was deposited in two layers to increase the number of connections between transistors, portions of which appear as flat, grayish areas beneath the conducting lines.

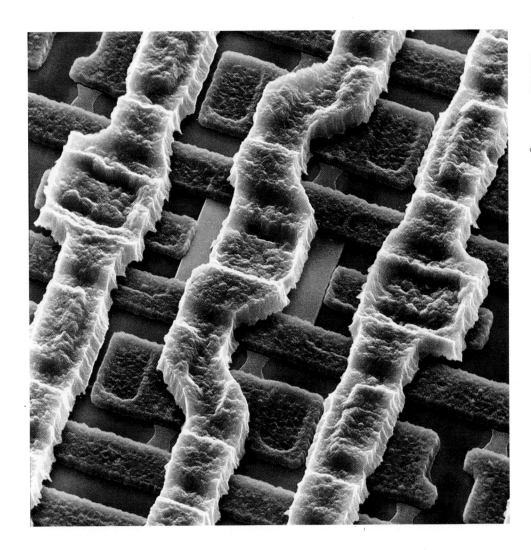

fell dramatically. Because all of a chip's elements are laid down at the same time, adding 10,000 components to a chip contributed negligibly to the cost of producing it. In 1971, for example, memory chips could store 256 bits of data at about two cents per bit. Two years later, the numbers were 1,024 bits and half a cent per bit. By the early 1980s, a memory chip could store more than 16,000 bits at a cost of about two thousandths of a cent per bit.

NEW SOLUTIONS TO NEW PROBLEMS

Changes in the semiconductor industry went beyond the factory floor. Chip designers, obligated to work in smaller dimensions with circuits of greater complexity, sometimes encountered unforeseen pitfalls. For example, the electrical resistance of a thin aluminum conducting path decreases faster than expected when the width of the conductor is narrowed. Higher resistance increased heat output more than anticipated. Not only does heat represent wasted electricity, but it also threatens the chips themselves, which fail more rapidly at high temperatures. To complicate matters, doped silicon is not a perfect conductor, since charge carriers, both electrons and holes, meet with some resistance caused by friction on an atomic scale as they move through the crystal lattice. Overcoming this resistance demands a small amount of power, which then is converted to heat. In addition, electricity can leak through thin insulating layers. When that happens, the chip malfunctions. Gradually, a set of design rules emerged that guided engineers around such problems, at least for the time being.

A new set of criteria for good circuit design also evolved. In the days of discrete components, electronics engineers found it rewarding to incorporate the maximum number of cheap resistors and capacitors and the fewest expensive transistors and diodes. But in an integrated circuit, capacitors and resistors, being less efficient made from silicon than their discrete-component cousins made from more ordinary materials, take up more space on the valuable silicon wafer than do transistors and diodes. As a consequence, designers devised new circuits that minimized—or even completely eliminated—the use of capacitors and resistors in favor of the less expensive transistors and diodes. For example, a transistor in which the base current is low could be used to limit current flow, just as a resistor would.

As chip designers tried to squeeze more elements onto a silicon wafer, the problems of power consumption and the resulting heat threatened to become insurmountable. Making circuit elements smaller reduces power consumption, but not enough to offset the increased demand for power by the ever-more-numerous elements that designers were adding to a chip. As power consumption rose, so did heat output, until circuits started to fail.

To counteract this problem, engineers added fans to computers or refrigerated their circuits to keep them cool. To chips they attached heat sinks, metallic radiators that, like the cooling fins of a lawn-mower engine, kept the temperature within safe limits. But those were stopgap measures. The cure would be a new type of transistor.

The bipolar transistor that engineers had favored throughout the 1960s came in two varieties, depending on whether the base was made of n-type or p-type silicon. In a p-n-p bipolar transistor, for example, emitter and collector regions of p-type silicon sandwich a base of n-type silicon. In an n-p-n transistor, the sequence is reversed. The n-p-n was the more common of the two because it was faster and easier to manufacture.

In the n-p-n transistor, the collector is attached to the positive side of a circuit, the emitter to the negative side. The attraction of the positive electrode causes the negatively charged electrons in the collector to gather there, and electrons in the emitter to gather at the emitter's junction with the base. For the electrons to pass through the base—that is, for a current to flow—the base voltage must be raised to pull the current from the emitter. Thus, applying a small positive voltage to the base turns the transistor on. A negative base voltage reduces the flow of electrons, turning the transistor off.

Bipolar transistors are lightning fast; some can switch between on and off in less than a billionth of a second. The main factor in determining a bipolar transistor's switching speed is the width of the base. A narrow base reduces the time it takes an electron to migrate from emitter to collector. But the price is high.

Power must be supplied to the transistor's base—and heat continues to build—whether the device is on or off. In addition, the complexities of building even the relatively simple, planar version of the bipolar transistor had become burdensome as the scale of integration increased. A simpler chip would be handy; one that consumed less power was a necessity. The answer would prove to be the metal-oxide-semiconductor, or MOS, transistor, named for its three major ingredients. The new device would not only consume less power, but it would occupy less space as well.

A ONE-WAY TRANSISTOR

In this kind of transistor only electrons or holes—but never both—carry the current from one electrode to the other. MOS transistors are thus known as unipolar devices. In the variety of MOS transistor known as an NMOS device, the current is carried by electrons. The transistor consists of two islands of n-type silicon embedded in a region of p-type silicon. Aluminum conductors connect directly to these islands. In a junction transistor, the islands would be called the emitter and collector; in a MOS transistor, one is called the source, the other the drain. The region of p-type silicon between source and drain is called the channel. A thin layer of insulating silicon dioxide lies on top of the channel, and a layer of metal, called the gate, is laid above the channel on the oxide, which isolates the aluminum electrically from the source and drain. The gate is analogous to the base of a junction transistor.

An NMOS transistor has the drain connected to the positive side of the circuit and the source connected to the negative side. In the absence of voltage to the gate, the transistor is off; the holes in the p-type channel prevent electrons from moving from the source to the drain. When a positive voltage is applied to the gate, an electric field is created that penetrates the oxide, repelling positively charged holes from the channel and attracting electrons. The result is a microscopically thin region of electron-rich silicon in the channel, which allows current to flow from source to drain. Because of the role that an electric field plays in the operation of MOS transistors, they are also known as MOS field-effect transistors, or MOSFETs.

Reversing the roles of p-type and n-type silicon results in a PMOS transistor, in which a negative voltage is applied to the gate and positively charged holes move from source to drain. NMOS transistors are slightly faster than PMOS (electrons move through the transistor faster than holes), but PMOS transistors are less expensive to manufacture because they are more tolerant than NMOS transistors of impurities in the silicon wafer.

The idea that current flow in semiconductors could be influenced by an electric field was an old one in this as-yet-immature technology. William Shockley, the Bell scientist credited with the invention of the transistor, had noted the phenomenon in 1948. However, the field effect was so capricious—weak in some experiments, strong in others—that it was regarded as little more than a curiosity at the time.

Further research indicated that the problem had to do with impurities in the silicon dioxide layer isolating the electrode that supplied the field. When the silicon dioxide was uncontaminated, the field effect was strong. However, the slightest amounts of foreign substances in the silicon dioxide reduced its insulating capabilities and allowed voltage to reach the underlying silicon, a circumstance that canceled or greatly diminished the field effect.

During the late 1950s, Bell physicist John Atalla headed a group of experimenters investigating methods of increasing the purity of silicon dioxide coatings for transistors, not so much with field-effect applications in mind as to improve the quality of the bipolar transistors of the day. Silicon dioxide having a predictably high degree of purity and insulating ability would make fabricating reliable transistors easier. Atalla's team found a way to get rid of contaminants in the silicon dioxide, a process that led him to suggest the idea of a field-effect transistor. In June 1960, at a conference of electronic engineers in Pittsburgh, Pennsylvania, Atalla and fellow physicist Dawon Khang demonstrated just such a device. Development of a practical MOS transistor took almost six years, as researchers struggled to create uniformly pure layers of silicon dioxide on one wafer after another of silicon.

A STAR PERFORMER

The results turned out to be worth the wait. Though MOS transistors are inherently slower than bipolar models (before any current can flow, an extra instant is required to pull electrons into the channel of a MOS transistor and turn it on), they more than compensate for the lost time in other ways. They consume very little power, for example, and correspondingly, they do not generate much heat. When a MOS transistor is off, no power is consumed at all, since the gate is kept at a potential of zero volts. In contrast, the base of a bipolar transistor always demands the application of voltage, and a computer made with bipolar transistors must employ a larger cooling system.

Because the power requirements and heat output of MOS transistors are so low, it is possible to fit more such transistors onto a piece of silicon. The newer devices are also simpler—and thus cheaper—to manufacture. To make a fairly common circuit using MOS transistors takes 40 percent fewer processing steps than building the circuit from bipolar transistors would require. Thus, when MOS transistors were introduced in the early 1970s, they further fueled higher levels of integration, enabling the semiconductor industry to continue following Moore's Law.

VARIATIONS ON A THEME

Several kinds of MOS technology soon became available to circuit designers, the most prevalent being CMOS, or complementary MOS. Such devices are made by combining NMOS and PMOS transistors on a single chip. One example is called an inverter, which turns ones into zeros and vice versa. Known also as a NOT gate *(pages 42-43)*, the inverter plays a crucial role in such fundamental computer operations as adding a pair of numbers. Although it is more complicated than standard MOS circuits—and thus more costly—the method of connecting the NMOS transistor to the PMOS transistor assures that CMOS-based logic circuitry draws less power than any other type in use, enabling computer manufacturers to build portable, battery-operated personal computers. The central microprocessors in the most advanced personal computers use CMOS circuits almost exclusively because of their low power consumption.

This small appetite for electricity leads to the biggest advantage of CMOS circuits: They generate even less heat than NMOS or PMOS transistors used alone. A typical microprocessor for a small computer built with NMOS transistors converts about 1.7 watts of electricity to heat. Adding memory chips and other circuits, also built with NMOS transistors, gives off nearly thirty times as much heat. In a microprocessor built with NMOS transistors, the chip temperatures will increase as much as fifty degrees Celsius, requiring heat sinks, vents, and fans. If the same microprocessor and memory chips were built with CMOS technology, power consumption would be 90 percent lower. As a result, chip temperatures would rise less than five degrees. No cooling provisions would be needed.

Lynn Conway leans on stacked copies of *Introduction to VLSI Systems,* the classic textbook on chip design that she coauthored with Carver Mead. Its straightforward presentation of rules and methods for creating complex integrated circuits helped demystify the science of designing chips.

DESIGNING THE CHIP

As the number of components in an integrated circuit rose, design possibilities proliferated. Designers gained new options and new limitations. In the 1970s, with the era of large-scale integration, a human designer had to keep track of the thousands of connections and components involved. But as the era of very large scale integration emerged, the demands on the chip designer grew enormously. They had to become Renaissance people of sorts. They had to learn the variations on standard design rules used by the companies they worked for. They were required to know fabrication techniques and the order in which the various processing steps are carried out. And they had to have marvelous powers of logic to perceive how hundreds of thousands, millions, even billions of components might fit onto a chip the size of a postage stamp. Little wonder that such people were rare.

At the beginning of the large-scale integration era, design was something of a black art, taught on the job to a few people by others who wore the mantle of master designer. The scarcity of designers and the length of time required to train them had a number of negative effects on the pace of technological advancement. Designing a chip was a slow process—the all-too-few designers available took years to create a relatively small number of new chips. In addition, enterprises that needed small quantities of a chip tailored to a special purpose were mostly out of luck. Custom-designed chips were economical only for applications that required thousands of them or for companies that were endowed with the resources to pay enormous engineering costs. A design bottleneck had developed.

CHIP-DESIGN VISIONARIES

The solution to the problem came from the work of two engineers, members of the chip-designer elite: Carver Mead, professor of electrical engineering at the California Institute of Technology, and Lynn Conway, now an engineering professor at the University of Michigan. Their collaboration sprouted in 1975, when Conway, who was then working for Xerox, began research for a joint Xerox-Caltech undertaking that came to be known as the Silicon Structures Project. "I was working on special-purpose architecture for image processing," recalled Conway, a transplanted New Yorker who studied electrical engineering at Columbia University. "I had become aware that there was a gap between the sorts of systems we could visualize and what we could actually get into hardware in a timely way." Mead, who has been at Caltech since his undergraduate days, calls himself a "lifer" there. The two engineers embarked on a search for ways to improve the hardware of highly integrated systems. Their conversations and ruminations on the topic led to a textbook that revolutionized the practice of integrated-circuit design.

A leading figure in the development of chip design, Carver Mead of the California Institute of Technology, has extended the frontiers of integrated-circuit technology with experimental chips that attempt to emulate powers of the human brain. Shown in the background are details from his design for a chip intended to mimic sight. Green squares are light sensors.

Introduction to VLSI Systems was the culmination of the Xerox-Caltech project. It set forth a number of techniques that serve as the foundation for creating any chip desired, no matter how complex. The authors developed rules for linking field-effect transistors into logical networks, thereby helping systems designers to visualize the mapping of circuitry. These rules are basic; they are formulated independently of fabrication limitations, such as the narrowest line that, when converted to aluminum in the finished chip, will conduct electricity as intended. The book also contained guidelines for how much silicon area a particular logic function should occupy. Conway and Mead skirted the issues of fine-tuning for fabrication by making fundamental assumptions about such questions as how wide metal should be in proportion to silicon.

While this information may not seem startling, it had previously been available to only a handful of designers who had held it almost entirely in their heads. The text demystified chip design and revolutionized the business of making integrated circuits. Suddenly, the wisdom and experience of the master chip designers was available from a library. "In electronics, a new wave comes through in bits and pieces," said Conway. "Usually after it has all evolved someone writes a book about it. What we decided to do was to write about it while it was still happening."

Even before the book was published in 1978, Conway loaded copies of the first chapters into her station wagon, carted them to the Massachusetts Institute of Technology in Boston, and used them to teach a course that became a legend. Mead taught a similar course at Caltech. The impact of their work was immediate. Before Conway and Mead, there were at most a handful of courses taught on LSI and VLSI circuit design. Soon after their chip-design bible appeared, nearly every major engineering school offered such a course, in which the students designed their own chips over a semester's time. Companies began holding seminars for the purpose of teaching their designers Conway and Mead's approach.

As a result, hundreds of well-trained designers entered the industry. In the past, only chipmaking companies, which produce enormous quantities of general-purpose chips, could afford to hire a staff of integrated-circuit designers. Now a wide variety of smaller businesses—manufacturers of products such as electronic keyboards, kitchen appliances, and even toys that depend on chips—can afford to employ qualified designers.

Furthermore, in an industry where all the manufacturing capacity was owned by the chipmakers, such as Fairchild, Intel, and Texas Instruments, the wealth of design talent and the chips they have created have spawned "semiconductor foundries," as Conway calls them, factories that will manufacture finished chips from a custom design.

GETTING COMPUTER DESIGN HELP

Though Conway and Mead made new rules for chip design, the steps involved in getting an integrated circuit to work are the same ones that have dominated the chipmaking industry since its earliest days. The type of transistors and the fabrication rules that are used at a company are that organization's chipmaking building blocks. Keeping them in mind, the designer arranges transistors, other components, and the interconnections between them in a way that will allow a chip to perform its intended function. The resulting logic schematic, as it is called, is then converted into the component patterns that will actually be used to make the chip. This design is tested to detect electrical and logic flaws. If it passes the test, the circuit is manufactured and sold.

In the 1960s, the logic schematic was created manually, with the chip designer preparing a rough circuit drawing as if using discrete components. A layout designer would then redraw the resulting diagram as patterns of silicon, insulation, and current-conducting strips of aluminum. Today, chip designers do very little of their work on paper. Instead, they use computer-aided design (CAD) programs to build their chips.

Typically, such a program contains a library of designs for silicon components—transistors, resistors, diodes, and capacitors—as well as the ability to move these components around the screen of a computer work station. In some instances, the library may contain complete subassemblies commonly used in chips, such as buffers for the temporary storage of data, flip-flops to use as memory cells, and multiplexers for increasing the volume of data that can be transmitted along a wire. With a combination of keystrokes and operations with a puck or a mouse—input devices that permit the designer to point at an object on the monitor and change its position—the designer selects the necessary elements for the chip and arranges them in patterns that will let the chip accomplish the tasks assigned to it. Connecting these elements on the screen tells the CAD software how they are intended to perform together, the basis for computerized testing of the design.

An even higher degree of computer assistance is achieved through the use of software called a silicon compiler. Like a compiler for a high-level computer language, which converts English-like statements of a program into ones and zeros for the computer, a silicon compiler translates a description of the functions that a chip designer wants an integrated circuit to perform—often in the form of a flow-chart-style diagram—into an actual layout of transistors and other components for the surface of a chip.

As helpful as computers can be in the planning and layout of integrated circuits, they have one major shortcoming. Compared to a human working manually, they are wasteful of space on a chip. The computer considers each function of an integrated circuit as a separate block of circuits, which are then pieced together on the surface of the chip. The individual blocks, however, rarely fit snugly, leaving gaps that a designer could use to further shrink the dimensions of the finished product.

Working with simulated input of high and low voltages, initiated by the designer and representing ones and zeros pulsing through the circuitry, the program calculates the correct output for the circuit arrangement specified in the layout. If the results agree with the designer's expectations, the circuit's logic is confirmed. If not, the design can be revised simply by moving components from one place to another on the screen.

Timing—that is, the coordination of chip activities—is a common source of difficulty. For example, if an operation in one part of a chip depends on information from an operation taking place in another area of the chip, and the first task gets under way before the data necessary for the second one arrives or after it has been replaced by something else, the chip will not function as hoped. Computer software inspects the chip circuitry for timing errors and notes them for the designer, who can usually solve the problem by delaying either the arrival of the data or the beginning of the task so that the two events are perfectly synchronized.

The CAD system checks further to be certain that all of its built-in chip-design rules have been observed: that none of the conductors is too narrow or positioned too close to another conductor; that insulating layers are thick and wide enough to do their job of preventing voltages from straying where they are not wanted; and that transistors are spaced far enough apart so that the operation of one does not influence the behavior of a neighbor.

When the design for the chip has passed all the tests, the chipmaking job is complete on the science front. Next, the action switches to the manufacturing front. Executing in silicon a design that comprises millions of transistors and millions of microscopically thin lines connecting them is arguably the ultimate challenge in chipmaking. Though much science and engineering has gone into developing the chip as an electrical device capable of manipulating huge volumes of data at blistering speed, an even larger effort has been invested in perfecting the fabrication techniques used to build chips.

Creating Patterns of Logic

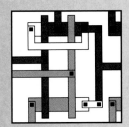

Designing a computer chip can be a mind-boggling task. As technological advances have made it possible to pack more and more components into smaller and smaller spaces, the work of uniting the many thousands of individual elements that make up a chip's circuitry has become increasingly difficult. And the stakes are high. A single error in the arrangement can render an entire chip worthless.

Mastering the many details of chip design can challenge even the most accomplished practitioner. Designers therefore approach the job by working through many different layers of abstraction. The process begins with an overall description of the chip's purpose. Then, step by step and component by component, the designer becomes more specific about how that purpose is to be achieved, eventually spelling out all the intricacies of transistor combinations and current flow.

Even the most basic chip component, such as the simple adder illustrated on the following pages, quickly becomes a maze of wiring connections and electronic switches. Chip architects have thus increasingly turned to computer-aided design processes that automate many of the steps that once solely depended on an individual's know-how and inspiration. These computerized techniques are becoming so sophisticated that soon computers will, in effect, be designing their own innards.

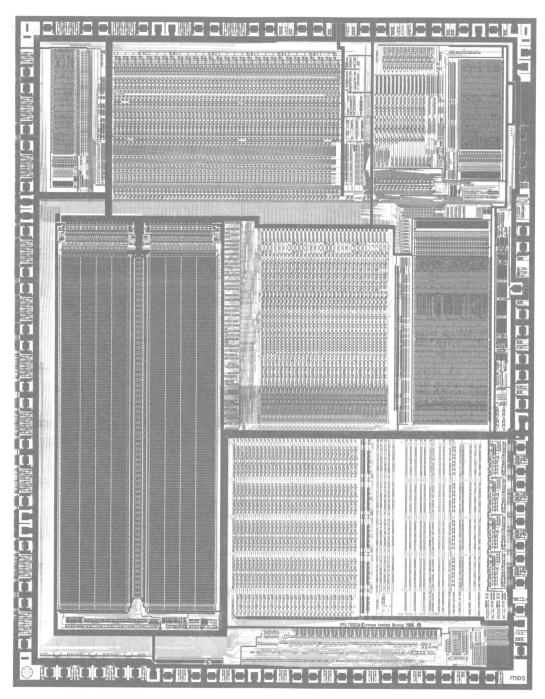

Where adding takes place. Nestled in the center of a microprocessor chip, a row of transistors and other circuitry *(box)* is organized to perform addition. Connections with other components, such as the two columns of memory cells to the left, facilitate the flow of input and output to and from the adder.

A Design Overview

The complex work of designing a chip begins with some relatively simple outlining of what the chip is expected to do and what components are needed to accomplish those tasks. The microprocessor chip at left, for example, incorporates the basic elements of a full-fledged computer, from memory cells to processing circuits. The designer's first chore is to identify these individual components and their functions so that they can be designed piece by piece.

An examination of one small portion of a microchip's processing circuitry, the adder, illustrates the concerns a designer must address in the initial planning stages. Ignoring for the moment the intricacies of how calculations will actually be performed, the designer starts by specifying the size of the numbers to be added and detailing the data channels that will carry inputs to the adder and outputs from it. The top drawing at right represents a simplified adder that will process numbers up to four digits long; in reality, most microchip adders handle larger figures, in some cases stretching to thirty-two places or even more.

A generalized layout such as this can be helpful in planning the interconnections between chip components, but it reveals nothing about how the adder itself will function. The designer therefore introduces more detail, based on the fact that multiple-digit numbers are added one place at a time from right to left, with carries from one calculation added to the next-highest-place column. As shown in the bottom drawing, the four-bit adder thus neatly breaks down into four one-bit adders corresponding to the four place columns of the numbers to be added; links between these individual adders provide for the carrying of values from one place to the next.

The designing process now becomes much more manageable. Instead of having to combine eight different input channels, the designer need only account for two input digits plus a carry—a much simpler structure that can easily be repeated as many times as necessary to make up an entire adder.

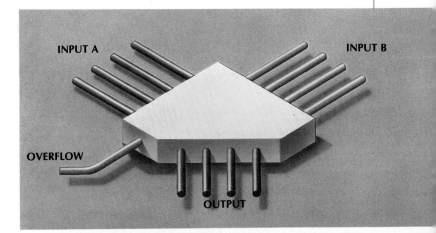

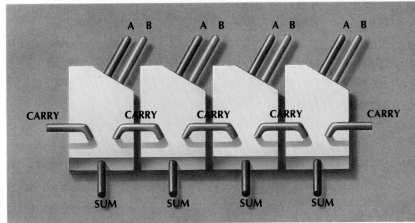

Outline for an adder. The drawings above delineate the basic input and output channels for a four-bit adder. The top drawing indicates that the adder will combine two four-digit numbers and produce as output a four-digit sum; the overflow channel handles any totals that exceed four places. The bottom drawing divides the adder into four identical stages, and digits are paired according to their place values, beginning at the right with the least-significant digits. Each stage yields a single sum and includes one channel for carrying in any value from a previous stage and another channel for carrying over any value to the next stage. Because the rightmost stage begins the process, its carry-in channel is permanently set at 0.

INPUT A	INPUT B	CARRY		SUM	CARRY
0	0	0		0	0
0	0	1		1	0
0	1	0		1	0
0	1	1		0	1
1	0	0		1	0
1	0	1		0	1
1	1	0		0	1
1	1	1		1	1

A table of possibilities. Listed above are the eight possible combinations of 0s and 1s that a one-bit adder might encounter when adding two input digits and a carry from a previous adder stage; the two columns to the right show the sum and carry produced by each combination of numbers. The last row, for example, indicates that adding three 1s produces an answer of 1, with a 1 to be carried out to the next adder stage.

Groundwork for a Binary Adder

Having successfully reduced the task at hand to the construction of a one-bit adder, the designer must now spell out in detail all the possible combinations of digits that the adder could receive as input, as well as the corresponding results it should produce as output. The fact that all the numbers are expressed in binary notation, in which every digit is either a 0 or a 1, helps limit the possibilities; in decimal notation, there would be ten options for each digit instead of two. Since the adder will handle three digits at a time—the two original inputs plus any carry from a previous adder's calculation—eight different combinations must be accounted for, as shown in the table at left.

The sums and carries that are listed in the two righthand columns of the table follow from the rules of binary addition, in which place columns increase by a power of two, rather than by a power of ten as in decimal arithmetic. Thus, whenever the inputs include two 1s, a 0 appears in the sum column and a 1 in the carry-out column, which represents the next-highest power of two.

The electronic circuits of a computer chip perform such arithmetic by means of organizational devices known as logic gates. These transform binary inputs according to different rules to produce logical—that is, predictable—results. Three basic types of logic gates used to organize computer circuits are shown at right. Clever combinations of these gates will yield the proper outputs for every possible combination of inputs fed to the adder.

NOT The simplest of all logic gates is the NOT gate, also known as an inverter. Unlike AND and OR gates, it accepts only one input, which it reverses, turning 0s into 1s and 1s into 0s.

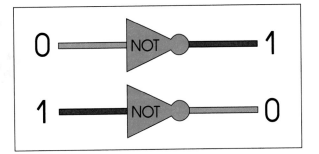

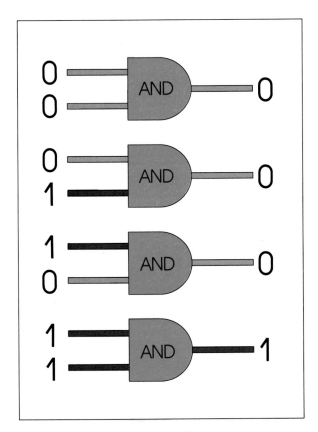

AND The four drawings above illustrate the basic operating principle of an AND gate: It yields a 1 only when all of its inputs are 1s. Although only two inputs are shown in each example, AND gates can actually accept any number of inputs, but like the other logic gates, they always deliver a single output.

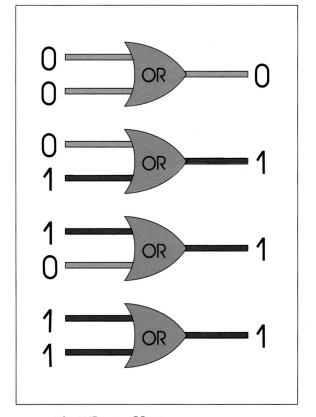

OR Like AND gates, OR gates accommodate two or more inputs and produce a single output, but they follow a different rule: The gate delivers a 1 whenever any of its inputs is a 1. Only if all of its inputs are 0s will the output also be 0.

INPUT A	INPUT B	CARRY	SUM	CARRY
0	0	0	0	0
0	0	1	1	0
0	1	0	1	0
0	1	1	0	1
1	0	0	1	0
1	0	1	0	1
1	1	0	0	1
1	1	1	1	1

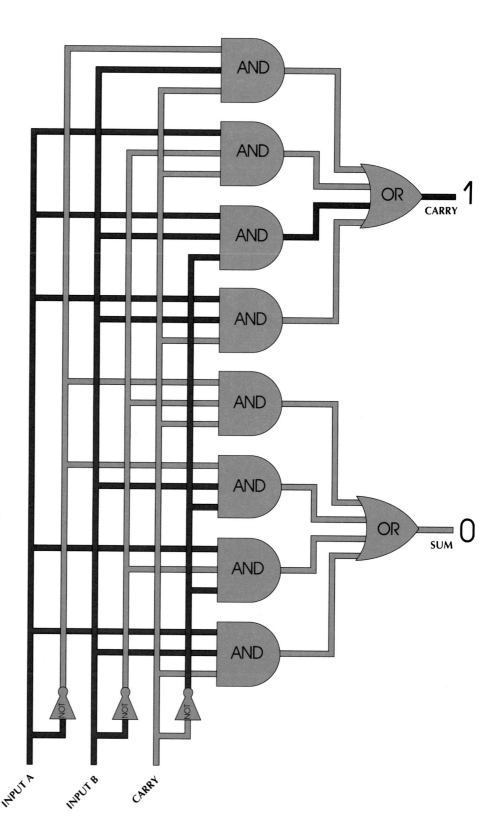

A route map for inputs. The network of pathways and logic gates at right is specially designed to produce the correct sum and carry for every combination of three inputs listed in the table above. Here, the network is shown processing the inputs from the table's seventh row, with dark purple representing a 1 and light purple a 0. Since each of the bottom four AND gates receives at least one 0 as input, each produces a 0 as output; these four 0s then become the inputs for an OR gate, which thus yields a 0 as output, representing the sum. But the third of the top four AND gates receives 1s for all of its inputs, so it passes a 1 to its OR gate, which in turn delivers an output of 1 as the carry.

Building a Logic Network

Because no single logic gate can function as an adder on its own, the chip designer must create a network of gates that will do the job. The gates are carefully organized so that the outputs of one set of gates become the inputs for another, until, at the end of the network, two outputs are generated that correspond to a sum and a carry.

Finding the most efficient combination of gates typically requires several stages of refinement. The designer begins with a methodical approach *(left)*, first dividing the network into two distinct sections—one to produce the sum and one the carry. Then, based on a systematic analysis of each row in the input-and-output table, the three inputs are channeled to four AND gates in each section, either passing directly to the gates or first being inverted by a NOT gate; the goal is to feed each AND gate a unique assortment of inputs so that no two gates duplicate each other's work. Finally, the outputs of the four AND gates are fed to a single OR gate, whose output represents the answer for that half of the network.

Such an arrangement functions perfectly for every combination of inputs in the table, but less complicated designs are also possible. The designer therefore searches for ways to achieve the same results with fewer gates *(below)*, often relying on an intuitive grasp of how logic gates work together. Each proposed version must then be carefully tested by tracing the flow of 1s and 0s through all the gates. For more complicated components than the one-bit adder represented here, the designer may employ an automated search program that follows prescribed rules for eliminating gates.

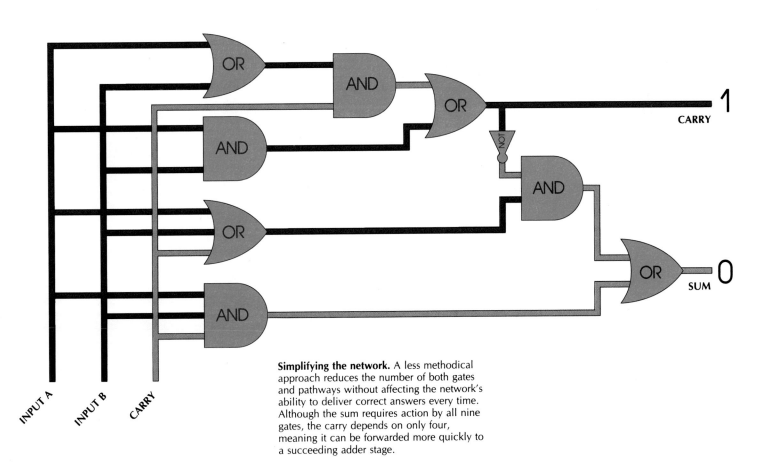

Simplifying the network. A less methodical approach reduces the number of both gates and pathways without affecting the network's ability to deliver correct answers every time. Although the sum requires action by all nine gates, the carry depends on only four, meaning it can be forwarded more quickly to a succeeding adder stage.

41

Silicon Switches for 0s and 1s

Because a logic network will eventually be translated into a series of electronic circuits, the designer's next job is to develop a layout of electronic components that will manipulate current in the same way that the network's gates handle 0s and 1s. The basic building blocks are transistors—on-off switches that control current flow through a circuit. The design must also account for the various conducting lines that will convey current to and from the transistors.

Chip circuits can be constructed from any of several types of semiconductor transistors. The CMOS (complementary metal-oxide semiconductor) circuits in this design involve two kinds of silicon transistors, indicated here by the colors yellow and green—denoting regions doped with impurities to make them either p-type or n-type (pages 84-85). These transistors respond in exactly opposite ways to signal voltages fed to them through a conducting line made of polycrystalline silicon, or polysilicon (red). A yellow, or p-type, transistor turns on in the presence of a low-voltage signal, which represents a 0; a green, or n-type, transistor is activated only by a high voltage, representing a 1. Switched-on transistors allow current to pass between metal conducting lines (blue).

By arranging doped regions, polysilicon and metal—greens, yellows, reds, and blues—in different ways, the designer can create the electrical equivalent of any type of logic gate. Shown on the opposite page is a design for a NOT gate, which requires only one pair of transistors and an associated branching line of polysilicon. The transistors are connected to parallel metal rails (blue) that conduct current charged by two steady levels of voltage whenever the power is turned on. The top rail carries a high voltage, while the bottom rail—which is actually the ground line for the circuit—represents a low voltage. Because one of the transistors in a pair will always be off when the other is on, either a high or a low voltage gets passed from the rails to other conducting lines to become the output for the logic gate.

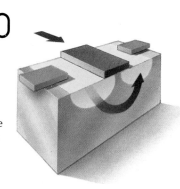

ON AND OFF WITH A 0

When a low voltage, representing a 0, enters the polysilicon line (red) of a p-type transistor, it enables current to flow between the two regions of p-type silicon (yellow), turning the switch on. Here, a high voltage (dark purple arrow) passes through.

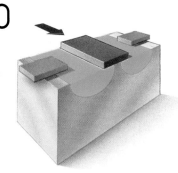

The same low voltage has no effect on an n-type transistor, so no current is able to pass between the two regions of n-type silicon (green), and the switch remains off.

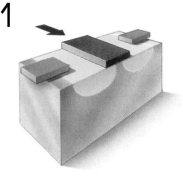

OFF AND ON WITH A 1

Because of the properties of p-type silicon, a high voltage—representing a 1—prevents current flow, keeping the p-type transistor turned off.

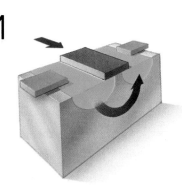

High voltage through the polysilicon line of an n-type transistor has the opposite effect, turning on the switch; in this case, a low voltage (light purple arrow) is forwarded.

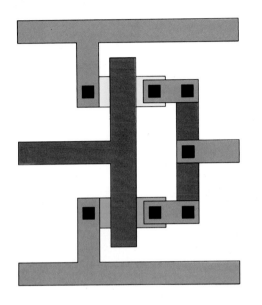

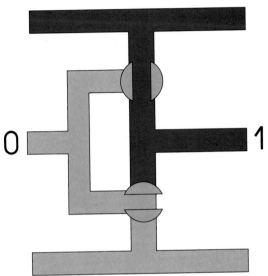

Processing a 0. At left is the designer's blueprint for the circuitry of a NOT gate. A branching line of polysilicon *(red)* will carry the signal current to both p-type *(yellow)* and n-type *(green)* transistors, which regulate the flow of either high or low voltage from two metal conducting rails *(blue)*. The simplified diagram at right shows the gate in action on a 0. Low-voltage input *(light purple)* turns on the p-type transistor and turns off the n-type. As a result, high voltage *(dark purple)* is delivered from the top rail to the output line, and the circuit forwards an output of 1.

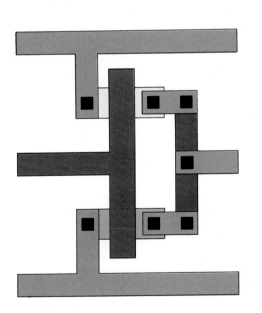

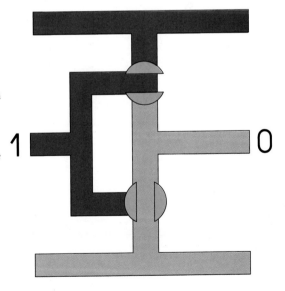

Processing a 1. The same circuit design *(left)* delivers a different result when the signal it receives as input is a 1. As shown at right, a high voltage *(dark purple)* through the polysilicon line turns off the p-type but turns on the n-type transistor. Thus the high voltage of the top rail is blocked, while low voltage *(light purple)* from the bottom rail is able to pass through, representing an output of 0.

Completing the Circuit Design

While designing a NOT gate, which accepts only one input, is relatively straightforward, AND and OR gates are more complicated. Every possible combination of inputs must be anticipated and the signal-conducting pathways and transistors of the circuit arranged to produce correct outputs. The two-input AND gate illustrated on these pages must be able to handle four different input combinations; three-input gates are even more complex.

Two types of transistor configurations—parallel and series—help duplicate in circuitry the rules of AND and OR gates. A parallel structure allows current to pass if any one of its transistors is turned on, just as an AND gate delivers a 0 if any of its inputs is 0, or an OR gate delivers a 1 if any of its inputs is a 1. A series structure covers the other possibility, conducting current only if all of its transistors are on; similarly, AND gates produce 1s only when fed all 1s, OR gates 0s only when fed all 0s.

Working one gate at a time, the designer uses these various options to create circuit patterns for every gate of the logic network. All the gates are then connected to form one complex adder circuit (pages 46-47), with conducting lines carefully placed to feed input signals to the proper gates.

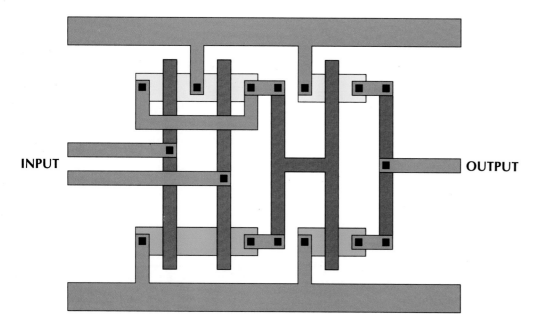

INPUT

OUTPUT

Design for an AND gate. The circuitry of an AND gate consists of two sets of transistors linked by an H-shaped polysilicon line *(red)*. The first set *(far left)* receives input signals through two vertical polysilicon lines, so that both the top and the bottom half contain two transistors apiece. The top two are parallel transistors; a U-shaped metal bridge ensures that if either transistor is on, current can flow from the top conducting rail to the polysilicon H. The bottom pair are series transistors. The output from the first set of transistors controls the second set *(near left)*, which functions as a NOT gate. In the presence of a low-voltage input, these transistors connect the output line to the high-voltage rail, producing a 1. A high voltage yields a 0.

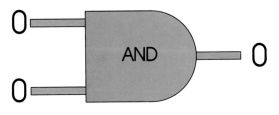

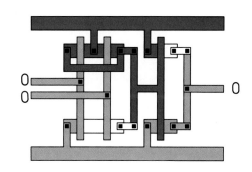

Two 0s turn on the top, parallel transistors *(right)*, passing a high voltage *(dark purple)* to the two single transistors. This turns on the bottom transistor, which conducts a low voltage *(light purple)*, or 0, as output.

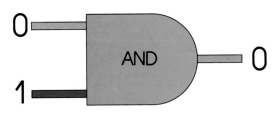

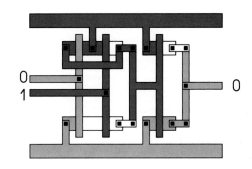

A combination of 0 and 1 as inputs still produces an output of 0; the left half of the parallel transistors is turned on by the top 0 and conducts high voltage through the U-shaped bridge and on to the single transistors.

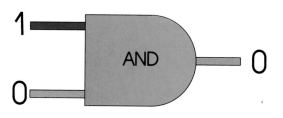

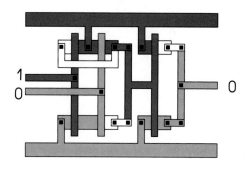

The input of a combination of 1 and 0 once again results in an output of 0, as the right half of the parallel transistors is turned on. The bottom, series transistors cannot pass current because both are not on.

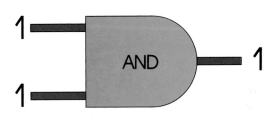

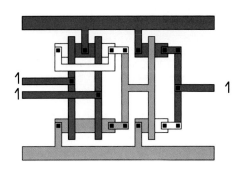

Combining only two 1s as input produces a 1 as output. Both parallel transistors are off, but both series transistors *(bottom)* are turned on, so that low voltage gets through and turns on the top, single transistor.

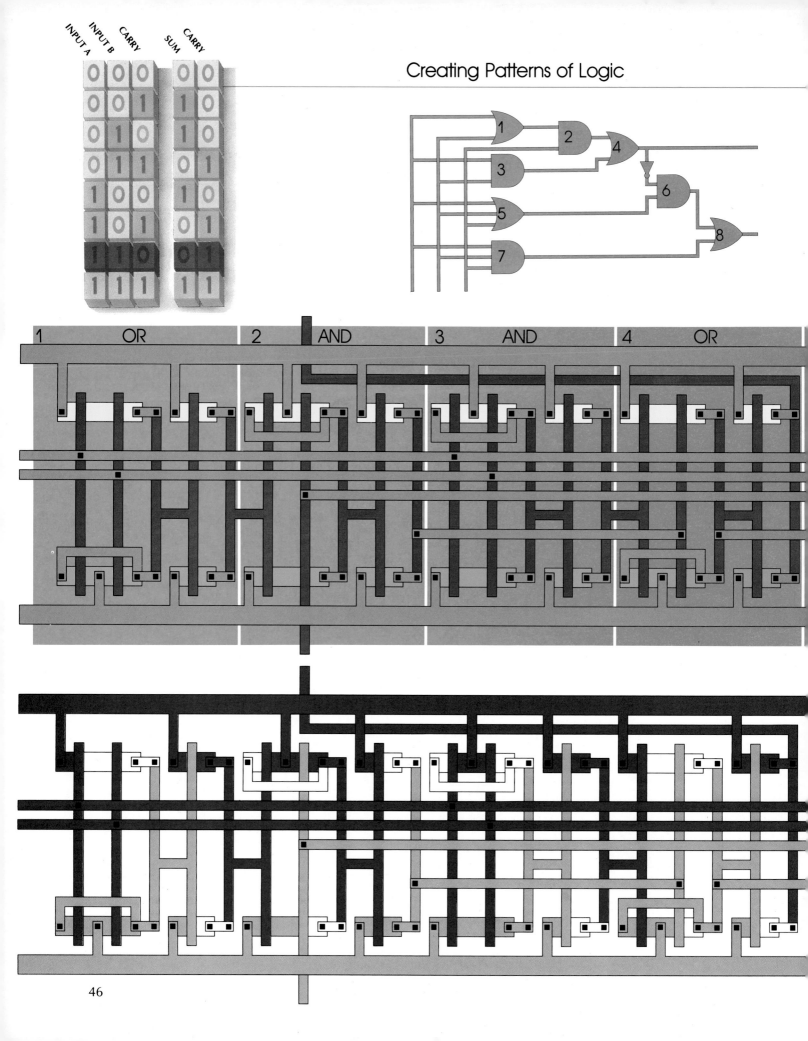

INPUT A INPUT B CARRY SUM CARRY

0	0	0	0	0
0	0	1	1	0
0	1	0	1	0
0	1	1	0	1
1	0	0	1	0
1	0	1	0	1
1	1	0	0	1
1	1	1	1	1

1	OR	2	AND	3	AND	4	OR

The completed adder. The top drawing below shows the adder's full circuit design, with the eight main gates arranged in the numerical sequence of the logic network at left; the lone NOT gate is created with a single metal connection between gates four and six. Contact points *(black dots)* on the central conducting lines ensure that inputs are forwarded to the correct gates. The carry input enters gate two from below, and the carry output is channeled back from gate four to align with it so separate adder circuits can be lined up on the chip. The bottom drawing shows the adder at work on the values in the seventh row of the input-output table.

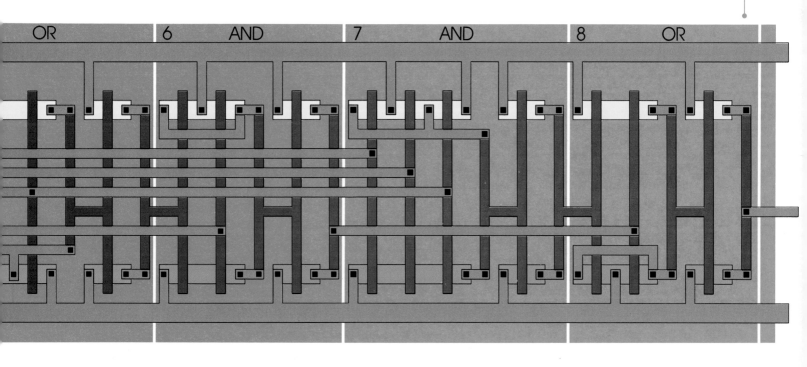

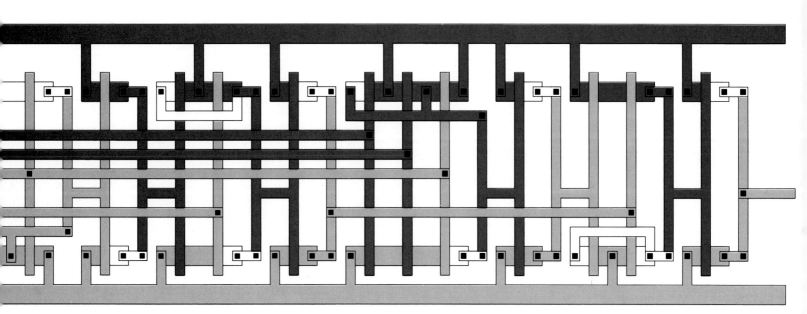

Patterned Masks to Implement the Design

With a completed circuit design for the adder in hand, the designer's final task is to create a series of masks, or stencil-like patterns, that correspond to the different types of material from which the actual circuit will be built. In essence, the design must be separated into as many parts as there are individual materials, with an additional pattern representing the contact points between materials. Each pattern is projected in turn onto silicon and various processes employed *(pages 75-87)* to imprint the patterns and then deposit, infuse, or remove materials until an entire working circuit is formed.

In the early days of chip manufacture, mask patterns were prepared and cut out by hand, then reduced photographically to the proper size. But advances in computer technology since then have continually improved the process. In one method, each mask design was first converted into digital form by being plotted on a special table covered with grid points. The data was then stored on disks or tape so that the pattern could be reproduced by machine.

The most modern computer-aided design systems make the creation of masks even easier. Working this way on a computer monitor, the designer fashions circuit patterns that are automatically stored; designs that were created previously can be recalled and combined at the touch of a button. The computer takes care of separating the design into its constituent parts and formats it for reproduction. The resulting design program can then be used to drive an electron beam that projects the pattern onto a chrome-covered glass plate, creating extremely precise masks.

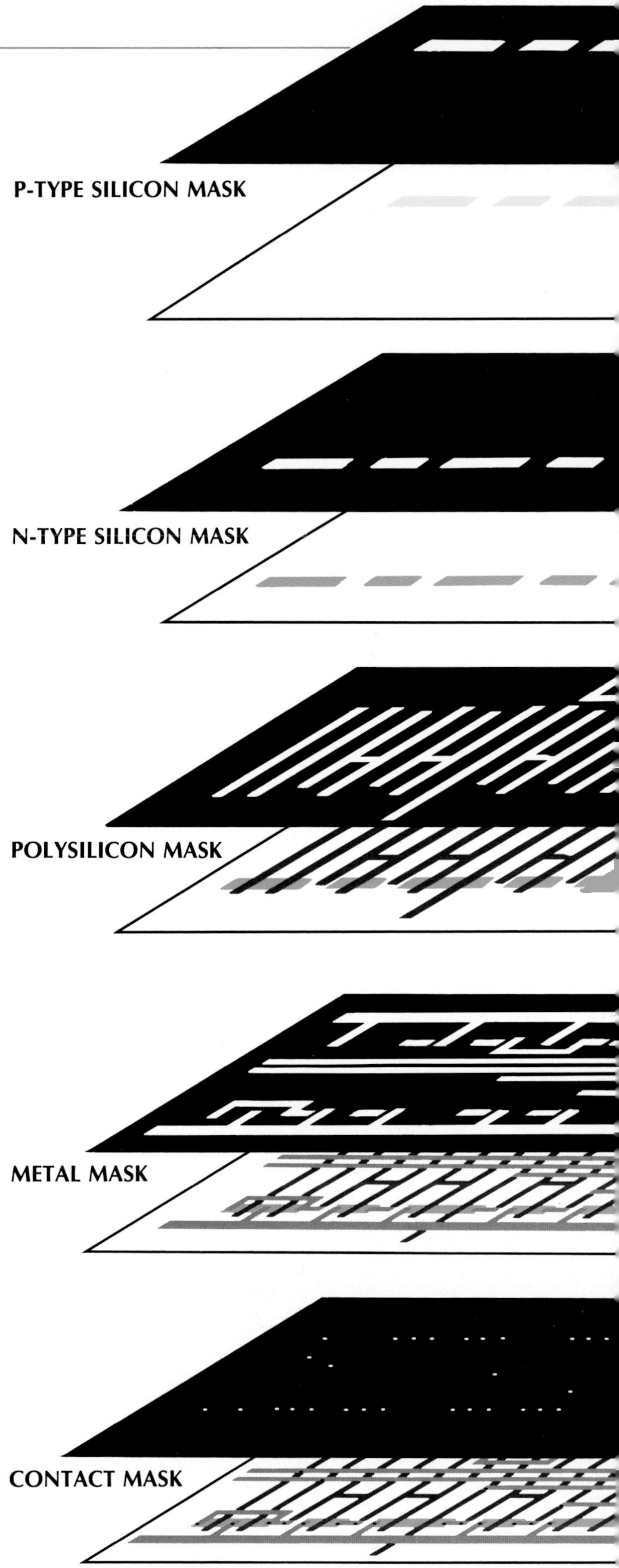

Imprinting a chip. Five separate masks are used one after the other to imprint circuit patterns on a silicon chip. Each mask outlines all the areas for one particular material, starting with the regions of p-type and n-type silicon, and ending with the contact points that provide connections between the metal and the other layers of the chip.

P-TYPE SILICON MASK

N-TYPE SILICON MASK

POLYSILICON MASK

METAL MASK

CONTACT MASK

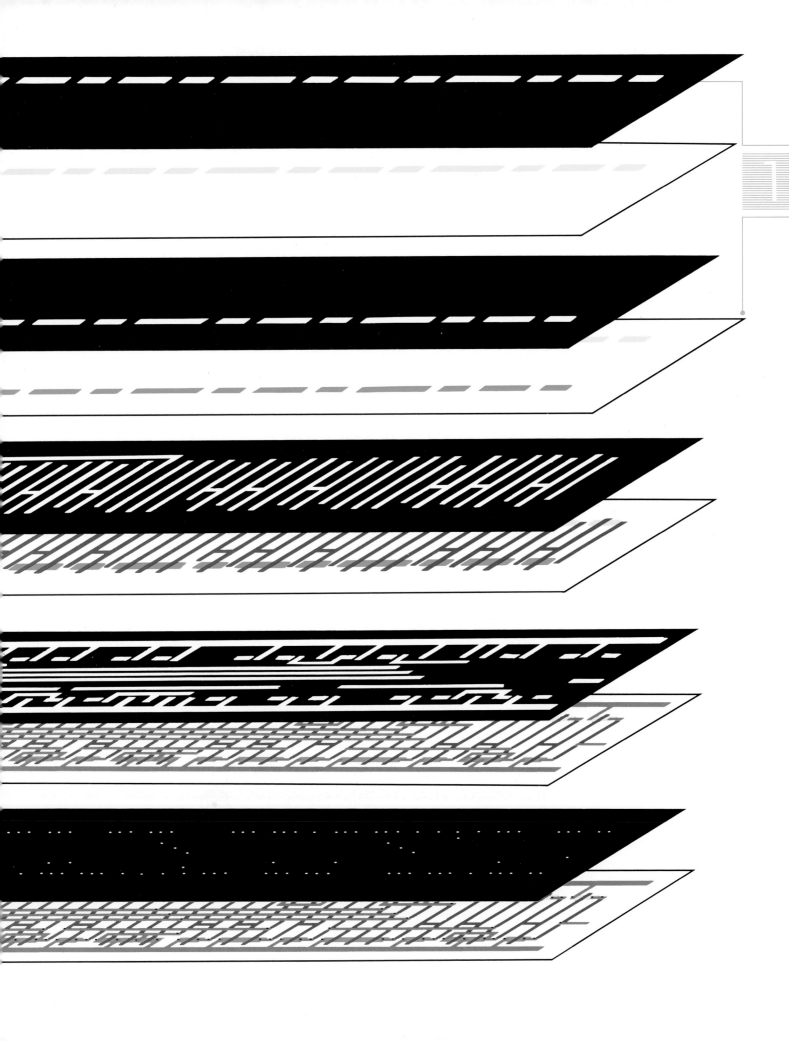

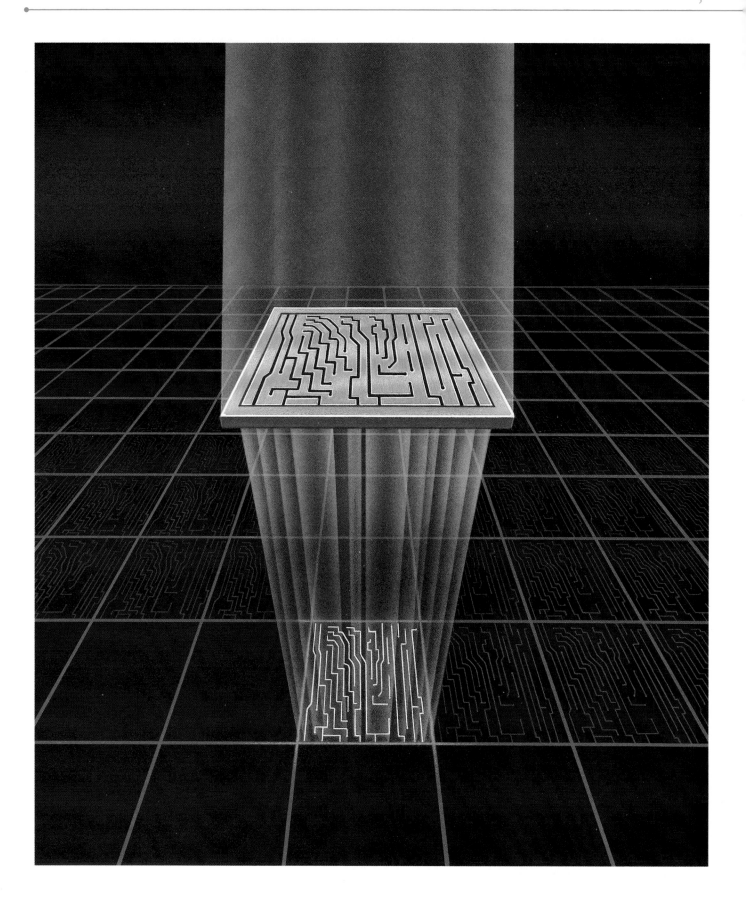

A Frontiering Industry

During a tour through a chipmaking plant at Intel, one of California's leading firms, a visitor got a firsthand glimpse of day-to-day reality on a semiconductor assembly line. In the so-called clean room, where silicon wafers are transformed into chips under scrupulously controlled conditions, a task force of young engineers was working frantically to figure out why the wafers in the latest batch were warping and cracking; the troubleshooting session would last throughout the night. In another area, production had been halted temporarily at a large silicon-processing instrument called an ion implanter, which now bore a hand-lettered sign announcing "Sorry, down for repair." It seems that a flustered new trainee had panicked and pressed the "stop" button earlier in the day, causing damage to the machine.

But the most telling detail was a wafer with red ink dots on each of its 200 chips, signifying that it had failed the final test and was bound for the trash bin. The reject was handed to the visitor with a wry explanation: At some point during one of the scores of production steps involved, a worker had hiccuped, jarring the wafer and ruining all 200 chips on it. With a market value of $10 per unit, that added up to a $2,000 hiccup.

This trio of routine glitches highlights some of the practical dimensions of the chipmaking industry. As Brookings Institution economist John Tilton has observed, building a new design prototype in the laboratory and manufacturing that same device in scale and volume and under competitive circumstances are two different things. "The former was really much easier," Tilton contends. "The people who made money really did the difficult task, which was producing these things in large quantities at very low prices."

Since the mid-1950s, the lion's share of glory has gone to the scientists and engineers who developed the semiconductor technology, literally doubling the microchip's complexity with every passing year. Their collective mission was, in essence, to keep pushing chip design to provide more circuit functions per second, at less cost per function and with improved reliability. For the manufacturers, this translated not only into finding new ways to pack ever-more electronic components onto smaller chips and maintain quality control, but duplicating the process over and over—up to hundreds of chips to a wafer, thousands of wafers a day.

The fledgling semiconductor industry faced a unique set of business problems as well. Its executives, who have typically been skilled in science rather than in business, now had to learn how to be recruiters, employers, managers, promoters, marketers, and comptrollers too. Fast-changing technology made the business highly unstable: Each increase in the number of elements packed onto an individual chip upped the ante a little more. Yet another strain was placed on the chipmakers' manufacturing methods and machinery, on the patience of their workers, and—the bottom line—on their ability to turn out enough usable chips to make a profit. At many a company, business pressures—including the mixed blessing of runaway growth—simply overwhelmed the managerial abilities of its leaders.

Nonetheless, the profit potential was great enough to attract scores of entrepreneurs. Beginning in the late 1960s, fierce competition for customers—with its attendant price wars and talent robbing—became another major survival factor. The most serious competition came, ultimately, from Japanese chipmakers, who caught up with the Americans in the late 1970s and continue to threaten the entire industry.

In this difficult environment, attrition has been a persistent reality—particularly in the industry's volatile early years. Of the seventy-one new firms that entered the semiconductor business between 1957 and 1976, records show that more than three-fourths were closed down or acquired by other companies. Conditions were more favorable to chipmakers over the subsequent decade in which 18 out of 157 start-ups went out of business or changed hands. Industry analysts attribute the increased success rate during that period to the development of new technologies, the ever-growing demand for microchips in products of all kinds, and the emergence of small, specialized "niche" markets that were ideally suited to small-scale business operations.

FROM PRUNE TREES TO CHIP FACTORIES

Much of this ongoing drama has been played out in a locale known as Silicon Valley, a thirty- by ten-mile strip lying mostly within the boundaries of Santa Clara County at the base of the San Francisco peninsula. The name has become synonymous with explosive growth and development, though Silicon Valley is not officially listed on any map of the area. Nor is it a valley as such—simply a flat area extending from the city of Palo Alto at the northwest end to the city of San Jose at the southeast.

At Silicon Valley's geographical center lies the community of Sunnyvale, home of the nation's greatest concentration of semiconductor manufacturers. In the early 1950s, Sunnyvale was a quiet little town that called itself the Prune Capital of the Nation. Many of its roadways were two-lane blacktops winding through orchards of plum, cherry, and apricot trees; manufacturing in the area consisted mainly of canneries and other food-processing plants.

Today, the fruit trees have virtually disappeared, replaced by the squat, boxy buildings that characterize the electronics industry. These structures are packed so tightly onto the landscape that an aerial view of Sunnyvale and the neighboring city of Santa Clara actually resembles a gigantic chip, its flat-roofed components laced with the circuitry of interstate highways. A 1982 Silicon Valley guidebook listed a total of more than 3,000 electronics companies and their affiliates that had set up shop in the area, creating new jobs at the rate of about 40,000 annually.

A kind of family tree links most of the seventy-some semiconductor companies currently operating in the San Francisco area. William Shockley, who co-invented the transistor at Bell Labs in the 1940s, is widely credited with sowing the seeds of Silicon Valley in 1955. That was the year he returned from the East Coast to his native town of Palo Alto to open his own transistor business—the area's first semiconductor firm.

In the long run, Shockley Semiconductor Laboratories produced little of significance, except for a group of gifted young engineers whom the eminent physicist had recruited from East Coast laboratories to work with him in Palo Alto—men who would soon strike out on their own to become leaders in their own right. When Shockley was awarded the Nobel Prize in 1956 for his work on the transistor, several colleagues gathered to toast him at a champagne breakfast; a snapshot taken on that occasion appears, in retrospect, like a photographic *Who's Who* of the early industry. Among the boyish figures sporting crew cuts and exuberant smiles are several members of the so-called Shockley Eight, who were to leave the company the following year to found Fairchild Semiconductor in the neighboring town of Mountain View. The celebrants included Robert Noyce, soon to be Fairchild's president, and Gordon Moore, who would team up with Noyce in 1968 to launch another industry giant, Intel. Other members of the Shockley Eight also eventually left the original group to form their own companies—such firms as Raytheon, Anelco, Signetics, Union Carbide Electronics, and Intersil.

Timing and talent made Fairchild a watershed phenomenon. In the early 1960s, amid the post-Sputnik emphasis on space and defense, Fairchild engineers persuaded the planners of the Minuteman intercontinental ballistic missile program to use their integrated circuits in place of the older, bulkier transistors that had served up to that time. This lucrative military business provided the springboard Noyce and his colleagues needed, and their firm dominated the industry for the remainder of the decade. Throughout the 1960s, it spawned

Colleagues toast William Shockley, seated at the head of the table, on the day he won the 1956 Nobel Prize in physics for his role in inventing the transistor. Among the guests are four of the Shockley Eight who, frustrated by their mentor's business methods, would soon leave to found Fairchild Semiconductor: Gordon Moore and Sheldon Roberts, seated at far left; Robert Noyce, standing at center, with glass; and Jay Last, far right.

offspring after offspring; twenty-seven new chipmaking firms in all had been launched by Fairchild alumni by 1969. At a conference held in Sunnyvale that year, all but two dozen of some 400 semiconductor people in attendance had at some point been Fairchild employees.

The mobility of that early era is still typical of the industry. "We all know each other because at one time or another everyone worked together," one Silicon Valley veteran told an interviewer recently. Added another: "You better watch out who you talk to and what you say to every person, because he may be your boss one day."

The pleasant climate and proximity of associated businesses prompted most new firms to set up shop right there in the San Francisco area. But as the industry outgrew tiny Santa Clara County, semiconductor enclaves sprang up elsewhere around the country. On the East Coast, start-up firms located primarily in the Boston suburbs and upstate New York. Other major concentrations have grown up around Colorado Springs, Phoenix—the home of Motorola—and Dallas, where Texas Instruments has its headquarters. The industry has nicknamed these centers Silicon Mountain, Silicon Desert, and Silicon Prairie, respectively. Japan has similar semiconductor congregations.

Although some large corporations such as IBM and Texas Instruments manufacture both the chips and the consumer products that use them, the majority of chipmaking firms make components and sell them to other companies. In the 1960s and 1970s, as integrated circuits became more powerful and were put to use in more and more different kinds of products, the chipmakers found themselves at the hub of an enormous universe of chip-dependent businesses. Their customers might range anywhere from the U.S. Defense Department and the aerospace and telecommunications industries to computer manufacturers and small electronics firms gobbling up chips to use in such things as video games, digital watches, and talking dolls.

SALES TAKES CENTER STAGE

As competition for these customers escalated, the marketing people who attracted and kept them became a bigger and bigger factor in a company's success. One of the more colorful Fairchild alumni is Jerry Sanders, an engineer-turned-marketer who showed Silicon Valley the value of shrewd salesmanship. Known for his aggressive tactics and his unabashedly lavish lifestyle, Sanders was hired away from Motorola to be marketing director of Fairchild at the age of thirty-one. A year later, in 1969, he left the company following a dispute and took seven colleagues with him to put together a new company, Advanced Micro Devices. The group's first challenge was to raise $1.5 million in venture capital—a struggle that very nearly put a premature end to their plans. As one of the partners recalls it, they were still several thousand dollars short of their goal just five minutes before the final deadline when a telephone call from an investor with $25,000 put them over the top.

The investment was well placed. At the outset, AMD aimed to fill a special niche in the industry: It was what is known as a second source company, a backup source of chips for customers who wanted to avoid total dependence on the primary seller. But with Sanders's flair for salesmanship (among other programs, he created price incentives for customers and insisted on producing chips that measured up to high-quality government standards, even though the company sold very few chips to the Defense Department), AMD became the fastest-growing chipmaker in the valley in the 1970s. Employee recruitment ads in 1979 featured the flamboyant president in a three-piece suit on a surfboard, boasting of $200 million sales figures and urging workers to join AMD's 8,000-strong staff and "Catch the Wave."

FIVE OUT OF TEN CHIPS

The chipmakers' vocabulary is one key to understanding their task, and one of the most important terms is yield—the ratio of good chips to the total produced. This percentage figure is the standard by which companies live or die. Although a company's actual yields are a closely held secret, manufacturers admit that yields below 50 percent are not uncommon, especially when a new chip design is introduced.

The potential yield of any given production run, of course, varies with the type of chip being made and the size of the original wafer, which is fabricated out of purified silicon by a separate process (pages 58-61). The smallest chips are simple logic chips, electronic circuits containing perhaps a dozen components, mostly transistors. More than 400 of these chips can fit onto a five-inch wafer. At the other end of the scale are the relatively large memory and microprocessor chips, packed with transistors, capacitors, and resistors (pages 68-69). A five-inch wafer can accommodate about 150 one-million-bit memory chips, or more than fifty of the larger thirty-two-bit microprocessors.

Whatever the maximum capacity of the wafer, the yield is always less than 100 percent because of inevitable defects in raw materials and the perils of contamination and damage as the chip passes through successive processing stages. If just 75 of 150 possible chips on a five-inch wafer are to be usable at the end of a ten-cycle fabrication run, for example, each cycle must operate at 93 percent of perfection. Small gains in each cycle produce large increases in final yield—and significant marketing advantages. A 95 percent success rate at each step in the example above will yield 90 usable chips, and a 97 percent rate will yield 110 good chips—almost half again as many as produced at the 93 percent level. And a 50 percent increase in chips brings in that much more sales income—an increase in revenue that far exceeds the production costs of increasing yield.

Because the microscopic particles that can kill a chip remain the same size regardless of the chip's dimensions, chipmakers have found a double advantage in smaller chips. In reducing the length of each component on a chip by half, the area shrinks to one quarter of its original size. This smaller chip has a one-quarter chance of catching a stray particle—and, of course, four times as many chips can be packed on the same size wafer at virtually no additional production cost. These simple mathematical facts have been the key to greater yields and higher profits.

For all its benefits, the trend toward miniaturization has produced added challenges for manufacturers. The chipmaking materials must be purer, the working environment cleaner, the machinery more accurate, and the quality checks more painstaking. Much of the strain of dealing with a fingernail-size product bearing hundreds of thousands, even millions of electronic components is borne by the employees themselves. Most of the assembly-line workers—the foot soldiers of the semiconductor industry—are women, of varied ages and ethnic backgrounds. The typical entry-level job pays only slightly more than the minimum wage. But although the work is often tedious and repetitious, many assembly-line tasks require considerable skill and judgment, so extensive training is required.

Because contamination from the environment remains Semiconductor Enemy Number One, workers must follow the most rigorous sanitizing procedures that have ever been devised. All employees wear special clothing in the clean room, entering and leaving through spaceshiplike air locks that route airborne contaminants outward, away from the fabrication areas. Makeup is forbidden for women, and men who have beards must wear special masks. Ordinary pencils, which produce dust, are banned; only ballpoint pens are allowed. With such daunting working conditions, simply keeping a trained staff on the job becomes a challenge; the employee turnover rate in the semiconductor industry—for both the assembly-line workers and the engineers—averages around 30 percent annually.

And with a product so tiny that hundreds of dollars worth can easily be slipped into a pocket or purse, theft inevitably enters the picture. Despite heavy security measures that are routinely taken, more than $20 million a year is lost in products that are stolen and sold on the "gray market." Silicon Valley's most notorious incident to date was the so-called Thanksgiving Theft of 1981, when 100 thirty-pound boxes of semiconductors worth more than $3 million disappeared over Thanksgiving weekend. The firm involved, Monolithic Memories, had stored the chips in a Sunnyvale warehouse, securing them in a tightly locked chicken-wire cage hooked up to cameras, motion detectors, and an electronic alarm system—and monitored by a security guard stationed fifty feet away. On Monday morning, the security setup was found intact, but the chips were gone. The security guard himself, along with two armed accomplices, was eventually convicted in the case.

A VOLATILE INDUSTRY

Of all the frustrations that come with the semiconductor territory, surely the most pervasive are the effects of constant change. Workers contend with the same learning curve that affects all manufacturing processes: With each new design there is first a struggle to master the logistics of making the device, accompanied by above-average defect rates and relatively high prices. Then, as experience grows and efficiency improves, yields rise and manufacturing costs decline. The price of a chip typically falls 20 to 30 percent for each doubling of production during the years of its life cycle, giving any firm that can progress rapidly on the learning curve a marketing edge over competitors.

But the volatile technology has consistently kept a step or two ahead of the manufacturers, which means that a product often becomes outdated just as the assembly-line routine gets up to speed—and so the cycle must start all over again with a new design. This constant balancing act makes the staffing aspects of the business difficult, to say the least. In the early days of Silicon Valley, for instance, one leading company improved an etching technique on one of its chips so dramatically that yields soared, and all orders for the device were met well ahead of schedule. This unexpected turn of events suddenly put 200 of the firm's employees out of work.

The instability in the business also complicates long-range marketing plans. Premium customers often demand forward-pricing deals—long-term contracts to pay low future prices for big advance orders, rather than current high prices. Errors in forecasting those future prices can mean substantial losses at billing time. And the short life span of semiconductor designs creates inventory problems on occasion: Surplus chips stored too long in a warehouse are likely to become obsolete before they can be sold.

THE ONE, TWO, THREE OF CHIPMAKING

Paradoxically, in this atmosphere of continuous change and evolution, the fundamentals of chipmaking have remained fairly constant *(pages 80-85)*. Fabrication still revolves around three basic operations—layering, patterning, and doping—which are repeated over and over during the assembly of the many layers composing a chip, somewhat like an elaborate silkscreening process. The designer's circuitry pattern for each individual layer is defined by a photographic imprinting procedure, over and over in identical rows of images. Etching removes unwanted surface material to leave the desired pattern in relief and expose underlying regions for further work. Specific areas of the chip are treated with dopants, impurities that enable that particular speck of silicon to conduct electricity in the proper way. And at each step, the active areas must be insulated from each other.

While most production methods have altered to some extent with the evolution of chipmaking technology, the insulating process has remained virtually unchanged since 1954, when two scientists at Bell Labs made a serendipitous discovery during routine experiments with silicon transistors. Up to that time, transistors were normally constructed of germanium, largely because early research at Purdue University had yielded good results with this material. Silicon, the principal ingredient of common sand, was a promising alternative, but its surface became rough and pitted when exposed to dopants at high temperatures. The Bell scientists, Carl Frosch and Lincoln Derick, were trying to solve the pitting problem by exposing silicon to various hot gases containing impurities.

One day, when Frosch was experimenting with a hydrogen-gas atmosphere, the hydrogen accidentally ignited, releasing water vapor into the system. This time, the scientists noted, the silicon emerged with a smooth surface; the oxidizing effects of the chemical reaction had coated the silicon with a thin, hard layer of silicon dioxide—glass. This layer proved impervious to most chemicals and to electrical charges; it was a perfect insulator and protector. When they tried the process on germanium, the resulting oxide was chemically active and electrically conductive. The implications were clear: Silicon could help manufacture its own insulator and protector; germanium could not. This small discovery was one of the industry's critical turning points; since that time, silicon has been the material of choice for semiconductors, and oxide insulating a key step in their manufacture.

Frosch and Derick also contributed to a second major chip-manufacturing process—etching away areas of insulation to gain selective access to layers beneath. In further transistor experiments, the two were reminded that silicon

Creating a Chip's Crystal Substrate

From the most rudimentary chips inside home appliances to the elaborate devices that run today's supercomputers, most integrated circuits are built upon wafers of silicon—business-card-thick disks between four and twelve inches in diameter. In principle, the manufacture of this foundation is simplicity itself: A large, cylindrical crystal of silicon is first grown, then sliced into thin disks, which are polished to a mirror finish. After a thorough cleaning, the wafers are sent to a chip-making facility, where each one is transformed into scores of integrated circuits.

The actual manufacture of wafers is a painstaking process. The first stage of wafer fabrication is the chemical purification of silicon found in common beach sand and elsewhere, a process that yields chunks of silicon that are 99.9999999 percent pure *(below)*. In this form, however, the silicon is useless for chipmaking. Not only are the chunks too irregular for slicing into wafers, but the material has a jumbled arrangement of atoms that impedes the flow of current.

These faults are overcome by converting the silicon chunks into a silicon crystal that has a latticelike atomic structure that permits electricity to flow in a predictable, controllable fashion. To accomplish the transformation, the chunks are melted in the crucible of a computer-controlled machine called a crystal puller. Trace amounts of chemicals such as arsenic or phosphorus are added to give the silicon electrical properties appropriate for the particular type of chip. A crystal of silicon lowered just below the surface of the melt becomes the seed for a huge silicon ingot that, once cool, is made into wafers.

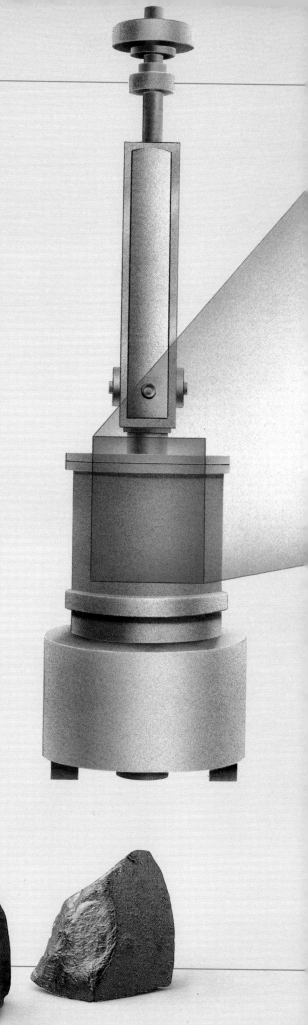

Seed Crystal

To form an ingot. The crystal puller first dips the tip of a pencil-shaped seed, rotating more slowly than a long-playing record, into a pool of molten silicon. Doing so cools the liquid around the seed just enough to cause a small amount of silicon to crystallize on the tip, extending the seed's lattice of atoms. The extension of the seed-crystal lattice continues as the crystal puller twirls the seed and lifts it away from the crucible, typically at the rate of two inches per hour or so for about thirty hours. The result is a roughly cylindrical ingot *(below).*

A silicon ingot. The photograph at right shows the lower end of a silicon ingot, reproduced actual size. This ingot averages a little more than five inches in diameter, is about fifty-four inches long and weighs more than seventy-five pounds. The taper at the tip is the result of a diminishing supply of molten silicon toward the end of the process.

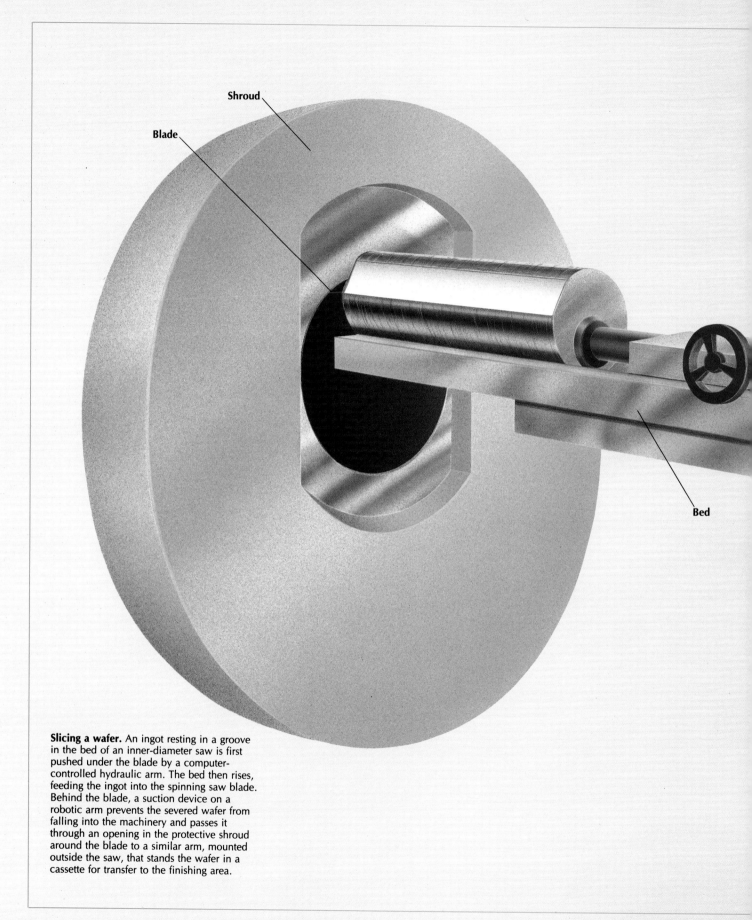

Shroud

Blade

Bed

Slicing a wafer. An ingot resting in a groove in the bed of an inner-diameter saw is first pushed under the blade by a computer-controlled hydraulic arm. The bed then rises, feeding the ingot into the spinning saw blade. Behind the blade, a suction device on a robotic arm prevents the severed wafer from falling into the machinery and passes it through an opening in the protective shroud around the blade to a similar arm, mounted outside the saw, that stands the wafer in a cassette for transfer to the finishing area.

Slicing an Ingot into Wafers

Converting an ingot into hundreds of wafers begins with a careful inspection of the crystal for size, purity, regularity in structure, and electrical characteristics. Undersized or defective sections are sawed off, as are the bottom and top. As much as half of the ingot may be rejected at this stage.

Next, the crystal is ground to the shape of a cylinder and probed with x-rays to determine the orientation of its crystal lattice. Then one side of the cylinder is ground flat along the ingot's entire length. This marker is used to align wafers during chip fabrication *(pages 75-87)*.

After the grinding is completed, the ingot is ready to be sliced with a special kind of circular saw known as an inner-diameter saw *(left)*. In this unusual machine, the blade is attached by its perimeter to a drive motor; in the center of the blade is a large opening, edged with diamond abrasives, that serves as the cutting edge. This design makes possible a thin yet rigid blade that minimizes the portion of an ingot ground into silicon dust.

The resulting wafers are stood on edge in slots of covered containers called cassettes and are taken to a finishing room. There, each wafer is first ground to a uniform thickness, then one side is polished to create a mirrorlike surface. Finally, the wafers are rinsed of lingering surface particles and residues in a series of chemical baths. The gleaming wafers are inspected, then packed in airtight containers for shipment to chip-fabrication plants.

Two views of a wafer. After preliminary finishing, a wafer has a dull surface *(below, left)* too rough for making into chips. Polishing, the next step, gives the wafer a glossy sheen *(below, right)*, making the surface smooth enough for the first circuit components to be added.

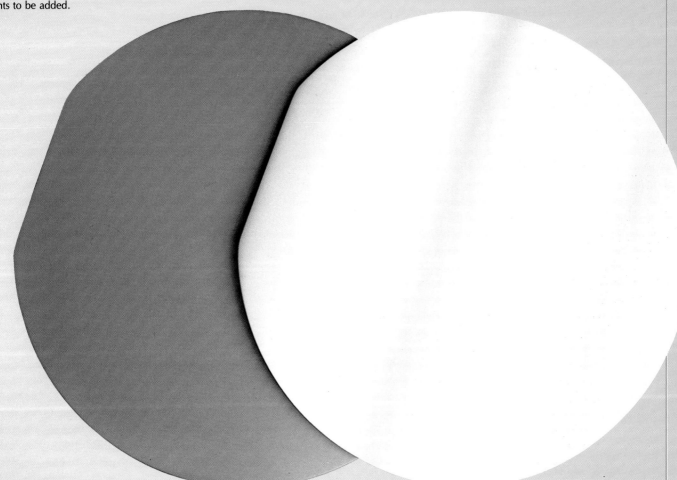

dioxide, like any other glass, can be dissolved by hydrofluoric acid—the familiar old scourge of high-school chemistry laboratories. The two scientists took advantage of this property to etch windows in the silicon dioxide layer, openings through which they could diffuse impurities into the silicon or insert wires to connect one point with another. Because the oxide was a good electrical insulator, it was also possible to lay down wires connecting several of these points without fear of a short circuit.

This process was adopted several years later in the development of the integrated circuit, repeated over and over again in order to create the interconnections on the surface of a chip. Originally, all etching was done by bathing the wafer in an acid or caustic chemical, as Frosch and Derick had done—so-called wet etching. The first generations of chipmakers realized that the wet etching was dissolving more of the chip than they intended—not only penetrating down vertically to a deeper layer, but also spreading out horizontally, across the surface of the chip. Vertical etching is desirable to remove the unwanted material; horizontal etching is not, because it erodes the edges of the line or feature that is supposed to remain.

In the early days of chip fabrication, this did not matter; the lines were wide enough to tolerate some horizontal etching. However, as the number of components packed on a chip increased and the dimensions of the individual features shrank to widths of only two or three micrometers, horizontal etching threatened to destroy vital features and damage chip performance. The solution was dry etching, using gases instead of liquids. Today, in a highly sophisticated variation called reactive ion etching, the wafer surface is bombarded with precisely directed ions—charged particles of the gas. The process is conducted within a vacuum, which has fewer air molecules to collide with and scatter the ions, and thus allows for their speedier and more precise placement on the wafer's surface.

New tools and procedures have also streamlined the process of adjusting the electrical conductivity of chips. The doping that was formerly done with high temperature gases is often accomplished now by firing high-energy ions of the dopant into the surface of the chip. Using a modern ion implanter—the type of machine that was inadvertently shut down by the unfortunate Intel trainee—a line operator can control the quantity and depth of doping more precisely and accomplish the task faster. And the wafer's exposure to heat is briefer, thus reducing the risk of warping or other damage.

FINE-TUNING A MACHINE

Of all aspects of chip fabrication, the photolithography used to define the circuit patterns has received more attention than any other. The workhorse of this process is a type of reduction camera known as a step-and-repeat camera, or stepper—named for the way it steps across a working surface, making exposure after exposure. Over the years, this machine has grown more precise, more sophisticated, and more laden with computer controls and laser measurements. And its very use has changed. Indeed, the story of the industry's continuous scramble to adapt its image-reproducing methods is a perfect illustration of the demands miniaturization and marketplace pressures have placed on chipmaking over the past two decades.

Initially, the stepper performed stage one of a two-stage operation: It was used to reproduce the pattern of an individual chip's circuitry over and over on a honeycomb-like stencil, or mask, the size of the finished wafer. First a large-scale image of the circuit design was created—by hand in the early days, later by computer—and photographically reduced to an image a fraction of its original size. This image, called a reticle, was still five to ten times larger than the final chip size. It was placed in the stepper, which reduced it to the size of the actual chip and repeatedly projected it onto the glass plate that would become the mask—one image for each chip on the wafer.

At the second stage, the mask pattern was transferred to a silicon wafer coated with a layer of light-sensitive material called a photoresist *(pages 82-83)*. Contact printing, the process learned by novice photographers, was the original technique used. Every time the mask contacted the coated wafer, however, both were slightly marred. At best, a hard-surface mask could be used for about a hundred exposures before dirt, pressure, and erosion began to degrade the reproduction quality. These problems were eliminated by moving the mask away from the wafer; one-tenth of a micrometer was enough. But even this tiny distance allowed light rays to spread, resulting in less exact lines on the wafer.

In the 1970s, when chip designers came up with line widths smaller than four or five micrometers, the manufacturers had to devise another solution—projection printing. A projection aligner, as the tool is called, worked much like a slide projector, using lenses to focus the light and direct the image from mask to wafer. Eventually, though, improved technology produced chips with features so fine that the wavelength of light became a critical factor. All waves—whether of light, water, or any other substance—have one property in common: They will not pass intact through an opening that is smaller than the length of the wave. When the width of the openings in a chipmaking mask became smaller than the wavelength of the light used in the aligner, the light broke up and scattered. The result was not a fine line, but a fuzzy one.

When the chipmakers tried using the shorter-length lightwaves at the ultraviolet end of the spectrum, they were stymied again, because glass lenses tend to absorb ultraviolet rays. Great energy could be put in at the lamp end, but it would come out in the form of heat, rather than light, since the lenses soaked up the short, nearly invisible rays. The solution to that new problem: Replace glass lenses with mirrors. And so, as the 1980s opened, the original light-refracting aligners gave way to reflective aligners, and a hall of mirrors rescued the chipmakers—for a while.

The next problem was posed not by designers, but by the economics of the business itself: the pressure to increase yield by using larger-diameter wafers. The larger the working area, the more difficult it is to build an optical system that will perform well across its entire field of view. With larger wafers, maintaining sharp focus became a particular problem: When the edges were in focus, the center was not. Minute variations in wafer thickness also resulted in a fluctuating depth of focus—the distance from the lens to the surface of the disk—between chips at various sites on the wafer. The aligners generally worked within variation tolerances of five to twelve micrometers, which meant that for acceptable results the wafer needed to be incredibly flat—equivalent to a road so level that it rises or falls only two-tenths of a centimeter in one mile.

For a while, the manufacturers tried an ingenious method of narrowing the aligner's aperture to a one-to-four-millimeter slit and shuttling mask and wafer past it on a movable carriage to expose only small portions at a time. This full-field scanner, as it was called, solved the lateral focus problem. But serious difficulties with small flaws in the mask, overlay accuracy, and variations in wafer thickness still remained.

Once again the conflict between scientific and business needs had brought the industry to an impasse. This time, in a rare exception to the rule, the problem was solved by going back to an older piece of technology: the step-and-repeat camera, which had been used to make masks for so many years. In theory, the idea was simple and direct: Move the stepper into the clean room where wafers are processed, and use the reticle, or stencil, to step the individual chip pattern right onto the silicon wafer, thereby eliminating the mask-making and transferring steps altogether.

In practice, it was not quite that easy. Engineers at GCA, the firm that manufactured the stepper, had considered a direct-stepping approach as early as 1960. But the machine was not designed for that use, and it needed several technical modifications to work properly. The notion remained on the back

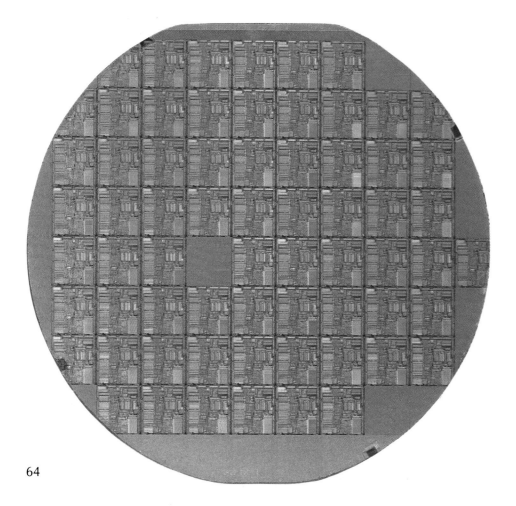

A single wafer of silicon, shown actual size, bears the imprint of dozens of identical microprocessor chips, each comprising more than 275,000 transistors and capable of rivaling the processing power of some mainframe computers. One square near the center of the wafer serves as a test chip; it consists of a few representative components that have been enlarged for easier access by testing probes. Partial chips overlapping the wafer's edge will be discarded.

burner until the mid-1970s, when Texas Instruments asked GCA to develop a direct stepper for use in making memory chips. The equipment firm finally tackled the adaptation project in earnest.

Like Cinderella preparing for the ball, the stepper inspired great hopes and enthusiasm in those who labored on the transformation. Their work essentially involved developing an autofocus system to compensate for minute changes in the distance between lens and wafer, devising a system for aligning the wafer accurately as the machine moved from chip to chip, and creating a stronger light source to reduce overall exposure time, since the images would now be stepped on the wafer one chip at a time.

With the adaptations complete, the wafer stepper was ready for action in 1977. Texas Instruments, ironically, changed its mind about the new machine it had helped to bring about, and placed no orders at the time. But the GCA team had faith in their creation. When they invited other manufacturers into the laboratory to observe the wafer stepper in operation, the machine quickly sold itself. "We were amazed at what we saw happen," recalls Jeanne Roussel, a GCA marketing manager who was an engineer on the project at the time. "We had been selling about twenty to twenty-five step-and-repeat cameras a year, and suddenly we were getting that many orders in one week. Let me tell you, those were very, very exciting times."

The wafer stepper turned out to be the breakthrough the industry had been waiting for, and it remains the standard method of handling photolithography today. It eliminates mask-making and its attendant errors, and presents a superior solution to the focus problems of large wafers. State-of-the-art computer controls precisely repeat the position of each chip for successive exposures, to provide accurate alignment of the various design layers. And in reducing the desired image as it is transferred to the wafer surface, the device reduces the size of flaws at the same time. Thus, a three-micrometer defect—enough to kill a modern chip—shrinks to a less-deadly 0.6-micrometer size when it is reduced in a 5X stepper. Although the stepper operates at about one-third the speed of a full-field scanning system, its greater accuracy, higher resolution, and its ability to expose 400 or more chips on a single eight-inch wafer compensate for the lower speed. The previous technology is now generally limited to smaller wafers and less-advanced designs.

A MATTER OF QUALITY

In a batch operation such as chipmaking, where errors have a disastrously cumulative effect, quality control becomes a major, ongoing part of the process. Since the first entrepreneurs set up shop to mass-produce semiconductors, rigorous monitoring and testing have been used to detect problems and weed out flawed chips at various checkpoints on the assembly line. The testing generally breaks down into two categories, physical and electrical. In the former, production workers physically test the wafers for defects; in the latter, they run tests to determine how the chips function. Among the dozens of items that must be checked are the thickness of the silicon dioxide insulating layer, the possibility that microscopic flaws on the mask or the reticle are being reproduced on the wafer, the accurate alignment of features, and the performance of individual circuit elements.

Automated equipment has begun to appear in the industry, but much of the actual inspection still requires a human being, a microscope, and a sharp eye. A technician determines the thickness of the wafer's silicon dioxide insulation, for example, using an ingenious visual test. The wafer is placed under a bright white light that strikes the surface and reflects back to the inspector's eye in repetitive bands of color, like a series of rainbows. The number of rainbows depends on the viewing angle, the refractive properties of the oxide, and the thickness of the oxide. Since the viewing angle is constant and the refractive properties of the oxide are known, the operator has only to refer to a chart at the inspection station to tell the thickness.

The size of features can be measured with great accuracy using a device known as a filar measuring eyepiece, a dimension-measuring instrument that is fitted to a microscope. However, its use requires aligning hairlines with edges of the features to be measured—an exacting, tiring task. By connecting the microscope to a video camera, the inspector can align the hairlines on the screen of a monitor, a far less tiring process. Originally, manual calculations were necessary to yield actual measurements of feature sizes; tool designers eventually provided inspectors with computer terminals capable of performing the calculations and storing the results for later analysis.

The more precise scrutiny required for modern chips is now done electronically, and with some irony: Machines that contain complex semiconductor chips are necessary to inspect semiconductor chips. Finding flaws in these crammed little devices is like "finding a matchstick in a five-acre field of hay," says Intel cofounder Gordon Moore. In 1984, some relief came when a "machine vision" company in Silicon Valley, KLA Instruments, produced the first fully automated chip inspection system. The industry saw KLA's Wafer Inspector as an unalloyed blessing. As one chipmaker remarked, "We were desperate for a machine that doesn't get tired and doesn't get fooled."

The Wafer Inspector has a special-purpose computer that stores in its memory a set of inspection criteria, along with a picture of what the perfect chip should look like. Fifty wafers are loaded into the machine at one time. The machine reads each wafer in a second, digitizes what it sees, and then uses special algorithms to compare the actual image with the ideal image in its memory. If the two images match, the wafer is sent on to further processing. If there appears to be an important discrepancy, the wafer is automatically shuttled off the production line.

THE AMERICAN WAY

During the early years of the industry, chipmakers' quality-control efforts were largely confined to this basic pass/fail process. American firms were satisfied if their products fell within the range of specifications set up by the designers. Wafers or chips that did not meet established standards were rejected—either sent back for reworking or tossed into the scrap bin. Whenever they could detect a pattern in the rejects, engineers made adjustments to keep the product within specifications. Along the way, they welcomed opportunities to improve the equipment and methods for inspecting wafers and chips as they moved down the assembly line. This was orthodox quality control in America—longstanding practice in the manufacture of everything from automobiles to candy bars—and the semiconductor pioneers saw no reason to suspect it would be inadequate in the electronic age.

Since the 1930s, an obscure American statistician named W. Edwards Deming had been preaching another theory of quality control, one that was more thoughtful and exacting—but far and away the road less traveled. Formally called statistical process control, the theory could be summed up in six simple words: Do it right the first time.

Applied to the semiconductor industry, Deming's ideas meant the American chipmakers were handling quality control all wrong. Merely finding bad products and throwing them out is no assurance of quality, Deming suggested; in fact, the reliance on this inspection-rejection system is a near-guarantee of an inferior—and expensive—product. It meant, in his words, that "a company is paying workers to make defects and then to correct them."

According to the principles of statistical process control, quality comes not from inspection, but from improvement of the process. To achieve this goal, manufacturers must maintain careful measurements of all aspects of the job, so they will understand the source and significance of irregularities; they must use the statistics gained by inspection and analysis of irregularities to improve the working materials and methods, and thus bring their products closer to the sought-after standard.

In the hurly-burly infancy of the semiconductor industry, these notions did not find much of an audience. After all, customers were accustomed to the limitations of vacuum tubes and fragile hand-wired connectors. Although every one of these components might work when it was installed, heat and vibration could soon generate a breakdown. But an integrated circuit kept on working; if it survived the first few hours, it would go on forever. What did it matter if the rate of flaws was high?

Manufacturers, too, had their hands full finding new ways to reach greater levels of integration. If yields were low at first, they would improve as workers gained experience with each new design. A few companies knew about Deming's theories and applied them; AT&T and IBM, both of whom manufactured chips for their own use, were the largest of these firms. But if the rest of Silicon Valley's chipmakers had heard of statistical process control—and most had not—they paid it no mind. They could afford to go their own way, until Japanese competition entered the picture.

CAUGHT BY SURPRISE

During the 1970s, American chipmakers were so busy mastering their trade that they paid little attention to Japan's growing potency in the field. Japanese electronics firms had already taken over the market in such consumer products as calculators, radios, and television sets. In the mid-1970s, with technology bought or borrowed from American firms, the Japanese moved into the semiconductor market with a vengeance. Less than a decade later, they had virtually walked away with the memory-chip business and were posing a serious threat to the entire industry.

The issue came to a dramatic head on March 25, 1980, at a meeting of the Electronics Industries Association in Washington, D.C. Members had gathered to hear a speech by Richard W. Anderson, general manager of the Data Systems

Building a Better Microprocessor

Of all the types of chips used in computers, the most complex is the microprocessor, whose web of circuitry incorporates all the main functions of a computer's central processing unit (CPU). Since its inception, the microprocessor has become an increasingly potent element in computerized devices, bringing greater processing power to everything from pocket calculators and automobile components to personal computers.

The first microprocessor (below), developed in 1971, was designed for a relatively simple desktop calculator. It could process only four bits of data at a time, severely restricting its speed. Large numbers, which require more than four bits for binary encoding, had to be handled in piecemeal fashion, first being divided into four-bit chunks and later recombined after processing by the chip's arithmetic logic unit (ALU). Because of the limited number of transistors, a sizable proportion of the chip was taken up by control circuits responsible for organizing the work, leaving less room for data-handling circuitry.

Over the years, improvements in manufacturing technology have enabled designers to pack more and more transistors onto a chip, greatly enhancing both the speed and the power of microprocessors. A modern thirty-two-bit chip can execute as many as seven million instructions per second, compared with 60,000 for the earlier four-bit chip. The secret lies in increased specialization of circuitry. In the chip at right, for example, separate sections devoted to memory management and bus control, as well as address and data buffers, maintain a highly efficient flow of data to and from the chip. Operating instructions, which on the first microprocessor were stored in the control circuitry alone, here reside as well in two ROM (read-only memory) sections and programmable logic arrays; frequently used data can be held in a temporary cache memory for more immediate access. And, instead of having the whole calculating burden fall on a single arithmetic logic unit, three ALUs share the work load.

Microprocessor progress. The first microprocessor ever built, the Intel 4004 (left), contained 2,250 transistors, with 16 wire connections (outer edge) for carrying signals to and from the chip. By contrast, the Motorola 68030 (right) includes about 300,000 transistors and 130 wire connections. Its circuitry is organized into distinct sections (right, bottom) that perform a variety of specialized functions, some of which previously would have required separate chips. An instruction pipeline feeds commands to three arithmetic logic units (ALUs), vastly improving processing speed. The actual size of each chip is indicated by the blue squares above and at far right.

ROM

Processing Control

Memory Management

Address Buffers

Instruction Pipeline

Programmable Logic Array

Data Buffers

Bus Control

Clock Generators   ALU   Cache Memory

Division of Hewlett-Packard, whose company was one of the world's largest purchasers of computer chips. What he had to say was simple: Hewlett-Packard would hereafter buy most of its chips from Japan rather than from domestic manufacturers. The reason he gave was equally simple: The quality of the United States-made chips fell far short of that of the Japanese products. Hewlett-Packard made the decision to buy Japanese, Anderson explained, after it had inspected 300,000 standard memory chips—sixteen-kilobit Random Access Memory devices (16K RAM in computer parlance)—half from Japan, half from the United States. The best American firms' chips had six times the failure rate of the worst Japanese producer.

Anderson's talk, later dubbed the Anderson Bombshell, was a public confirmation of a long-held private suspicion, but that made it no less dismaying. The people in the room had invented the chip; they had developed the ingenious, exacting process that turns common beach sand into the miraculous instrument of the computer revolution. Yet they were losing business to engineers who—in their view—had merely followed their lead. The American manufacturers were crying foul, complaining that their Japanese competitors enjoyed the advantages of heavy government subsidies, cheap capital, and protective foreign trade policies.

These charges were true. But on that March day in 1980, there was no longer any denying that the real Japanese secret weapon in the chip wars, as they were called, was quality. The American companies clearly were doing something wrong. Although they had taught the world how to put thousands of electronic circuits on a tiny sliver of silicon, they obviously had not themselves learned how to do it well.

In a burst of catch-up activity that was almost painfully ironic, American firms set about discovering the Japanese formula for success. Seminars on Japanese quality control were held; books on Japanese management methods were read; investigative teams went to Tokyo to observe the Japanese in action. What the Americans found was an approach to quality control that was heavily preventive, one that began with the planning of the production line to reduce the probability of defects occurring in the first place.

The Japanese, too, were inspecting their products and gathering and analyzing data about the fabrication process. However, they were using those statistics in a different way. They considered the stream of information to be part of the process, not incidental to it, and they fed the data back into the system so they could improve the fundamental operation. They were, in short, using statistical process control—and they had learned it firsthand from no less a teacher than W. Edwards Deming.

A PROPHET FROM THE MIDWEST

The story of Deming's watershed role in the chip wars is one of dogged conviction and coincidence, of a war-torn nation's eagerness to rebuild, and of the ultimate vindication of a man who was for half a century a prophet without honor in his own country. Born in 1900 in Sioux City, Iowa, Deming was raised on a Wyoming homestead, where he displayed early talents in the areas of math, science, and music. He eventually earned degrees in mathematics and physics, including a doctorate from Yale University, and began a long career as a government scientist, first with the U.S. Department of Agriculture, and later at the Census Bureau.

While he was at Agriculture, Deming was introduced to a Bell Labs statistician named Walter Shewhart, who had first developed the techniques of statistical process control in the early 1920s. The younger man became intrigued with Shewhart's theories, which involved defining the limits of random variation in a worker's task so that results outside those limits could be identified and the causes examined. For several years, Deming traveled regularly to New York to study with Shewhart.

It was a propitious time for statisticians in government. As the 1940 census approached, a debate raged over the use of statistical sampling of the population. Sampling is now a common technique, used not only in national censuses but also in poll taking of all kinds, from the Nielsen surveys of television viewing to political candidates' voter-preference polls. But before 1940, census takers had polled every individual, a time-consuming process in itself that led to additional expense and time in tabulating the results. Partisans of sampling maintained that statistical techniques could yield accurate information cheaper and faster—and that sampling would permit interviewers to ask more questions and thus expand the information that would be available to the government. Census Bureau officials decided to try the technique, and Deming was placed in charge of the sampling program.

The new assignment also offered Deming a chance to apply what he had learned from Shewhart to an important process-control task. Census information was recorded on punch cards, every single one of which had to be checked for errors. Despite this precaution, errors still crept through. Deming examined the system, analyzed the errors, and instituted an operator-training program that so reduced the error rate that only one in three cards required checking to achieve almost the same results as 100 percent inspection. Within a short time, this knowledge and experience would be put to other use, manufacturing matériel for the military in World War II. During the war, Deming worked with General Electric, the Hoover Corporation, Stanford University, and other organizations in a campaign that ultimately taught Shewhart's statistical methods to more than 30,000 students.

After World War II, Deming left government service and went into business for himself as a statistical consultant, but American manufacturers no longer had much use for his philosophy of statistical process control. Domestic industry was booming, and with the resurgent economy, consumer demand was such that companies could sell almost anything they could make. Manufacturing was done on classical assembly lines, and the work was managed by ardent followers of Frederick Winslow Taylor, who took a stopwatch into the workplace and placed efficiency—the production of the greatest volume at the lowest cost—above all other goals. Quality control was incidental.

The situation was far less rosy for manufacturers in Japan, which had surrendered to the Allies with its industrial base destroyed. To markets around the world, "Made in Japan" meant junk. Deming first stepped into this bleak situation in 1947, when he was invited to Tokyo by General Douglas MacArthur to help the occupation forces with the first postwar census. Instead of spending his time in the company of the large American colony, Deming went off on his own to meet and cultivate the Japanese who were struggling to rebuild their nation's industry.

A Japanese statistician named E. E. Nishibori, familiar with Deming's work, found out by chance that the American was working in Japan. Nishibori tracked Deming down and invited him to speak at a quality-control workshop for the Union of Japanese Scientists and Engineers. That speech led to others; Japanese scientists were evidently impressed with the theories that American industry had flirted with and then dismissed in the postwar business boom. In July of 1950, Deming agreed to return and appear free—on the condition that the Japanese would arrange for him to speak to the nation's top business executives. Experience had taught Deming that, unless top management was behind the quality effort, it was doomed to fail.

Twenty-one industry leaders showed up to hear that landmark address. Deming told them how they could capture world markets in five years, and then he gratified them with an astonishing prophecy. "I predicted," he recalled in a recent interview, "that Japanese manufacturers would come to dominate world markets and have their competitors crying for protection. At that time I was the only person in the world who believed that. But it was so simple. You could see that this society was receptive to the ideas that are necessary for quality and productivity."

The Japanese executives, having nothing to lose, embraced statistical process control, applying it in the next few years to the manufacture of automobiles, motorcycles, cameras, stereos, televisions, office products, and—eventually—semiconductors. Although many factors contributed to their success story, the Japanese awarded Deming a hero's share of the credit. He is the only living American who holds Japan's premier imperial honor, the Second Order of the Sacred Treasure. Moreover, since the early 1950s, a prestigious award named the Demingu Sho, or Deming Prize, has been presented Oscar-style each year for achievement in the field of statistical process control.

A CONTINUOUS PROCESS

The genius of statistical process control is that it shows businesses when to act and when to leave things alone. If the quality of a product is judged solely by the specifications, neither the producer nor the customer has any idea where the product falls within the range of specifications. It is like not knowing whether the student's passing grade is a 60 or a 90. A high-quality product, Shewhart and Deming insisted, is not only one that meets high standards of performance—the specifications—but also one that exhibits very little variation from one sample to the next. Their logical solution for businesses: Don't settle for a product that simply meets specifications; instead, strive to improve the product by reducing variations. And this goal can be achieved only by constantly studying the fabrication process and applying the fruits of that study—statistical pointers—to improvement of the process.

Through painstaking analysis of manufacturing operations, the two statisticians calculated that a scant six percent of all defects can be attributed to worker errors. Deming called these "special causes," the kind of one-time problem that can be corrected by adjusting a machine or giving a pep talk. The other 94 percent of defects proceed from "common causes" built into the system—poor maintenance, poor training, and other chronic matters immune to worker solutions. Thus Deming reached his conclusion about the critical role of management's commitment in bringing about quality.

Because common causes tend to fall within what statisticians call a range of normal variation, management can use statistical analysis to determine which defects are caused by special circumstances, which ones are faults of the system, and where a fault of the latter type falls within the variation range. The firm can then take remedial measures that are appropriate to the particular problem, rather than make a scattershot fix-it attempt that might only end up making matters worse. Such a system then, governed by statistical information, is said to be "in control."

These were some of the lessons that Japanese managers learned in 1950, and that American chipmakers were forced to learn in the 1980s. Setting quality in their sights, they bought Deming's books and studied the techniques of statistical process control. The statistician himself, by now an octogenarian but still going strong, found his appointment book filled with speaking engagements; major American firms were suddenly paying him fees upward of $10,000 a day to teach them what he had taught the Japanese thirty years earlier for nothing.

The system they were adopting calls for constant data gathering and analysis—just the kind of work computers are made for. Measurements taken with the filar microscope, for example, are gathered and analyzed by the computer that calculates them, and the results are directed to the fabrication section responsible. Results of visual tests—like those for oxide thickness—are also entered in a computer for analysis and review. The automated Wafer Inspector feeds data to a computer add-on that analyzes and displays it in graphic form, so it can be more easily understood and acted upon. The data gathered during inspection is turned over to process-control engineers, who use it to help spot trends and direct their attention to solving the most damaging problems. Thus, the very product of the fabrication process is playing a large and growing role in assuring its quality.

Two years after delivering his bombshell in Washington, Anderson reported that the Americans had largely achieved their quality-improvement goals. But boom-or-bust cycles pegged to the overall economy continued to buffet the semiconductor business throughout the decade, and the chip wars with Japan have raged on unabated as well. By the late 1980s, U.S. chipmakers' share of the world market had slipped from 60 to just 43 percent, while Japan's share rose to about 44 percent. As a survival strategy, domestic chipmakers have begun to imitate two other Japanese practices: closer ties with government and greater cooperation within the industry. In 1981, Robert Noyce of Intel and AMD's Jerry Sanders reversed Silicon Valley's longstanding traditions of lone-wolf entrepreneurism and rivalry by agreeing to collaborate on the design and production of new products.

The underlying principle—that twice as much research-and-development benefit could be gained by pooling information rather than jealously guarding it—has been carried out on a larger scale in recent years at several joint research-and-development centers around the country. At one such consortium, North Carolina's Semiconductor Research Corporation, vice president Robert Burger said that erstwhile rivals are often surprised and amused when they sit down together face to face to share information. "In a lot of cases, they discover that what they thought was their own private corporate secret was the same big secret all the other companies were keeping." Some firms have simply pooled their resources by merging. In 1987, four notable Silicon Valley veterans became two: AMD merged with Monolithic Memories, and the venerable Fairchild Semiconductor was acquired by National Semiconductor, one of its early offspring on the family tree.

American chipmakers, aware of the beneficial alliance between Japan's government and its electronics industry, are also bidding—some of them grudgingly—for help from the federal government. Spurred by the critical role semiconductors play in national security applications, a new venture called Sematech, jointly financed by the industry and the Defense Department, was established in 1987 to support American chipmakers. Industry lobbyists have also stepped up efforts to win government funding for research, tax laws that would favor investment in new technology and equipment, and more advantageous trade policies. Advocates of such government intervention see it as the most effective way to prevent an industry shakeout that would leave only giants such as IBM and AT&T still in business.

True to form, these challenges on the business front came at a time when the chipmakers were juggling new breakthroughs in technology: They were about to deliver a chip capable of storing a million bits of information. And already the question was asked, "When will it be a billion?"

To convey some idea of its complexity and power, the circuitry of a microchip has often been compared to a street map—one with electrons rather than vehicles moving along its byways. Thus, the first chips of the mid-1960s resembled the street layout of a small town, and the microprocessors of the early 1980s swelled to the complexity of a huge metropolitan area. By the same analogy, the billion-bit chips envisioned for the future would be nearly unfathomable in their intricacy—equivalent to a street map covering the entire continent of North America.

Challenges
of Fabrication

2

Engraving the Lord's Prayer on the head of a pin is a bumble-fingered job compared to the task of making the tiny chips that lie at the hearts of computers, cameras, videocassette recorders, and all the other electronic marvels of the age. To squeeze thousands or even millions of circuit components—resistors, capacitors, diodes, as well as transistors—into a half-inch square of silicon is perhaps the world's most demanding craft and the least forgiving of imperfection. Because a chip can be ruined by a single speck of dust too small to see, chipmakers are obsessed with cleanliness. And since fabrication processes involve tolerances smaller than the wavelength of light and are sensitive to the faintest shake from a footstep, the workplaces must be uniquely designed to isolate vibration.

Yet chipmaking is fundamentally simple. Its basic technique is the century-old process of photolithography, used to print newspapers, magazines, and books (including this one). But this common craft has become incredibly elaborate. Instead of reproducing type and pictures in two dimensions on a sheet of paper, it shapes layers of silicon and other materials in succession so that each layer can be treated and interconnected to perform the electronic functions of a three-dimensional integrated circuit.

These methods, demanding as they are of precision, nonetheless achieve a new level of mass production. A small plant, employing possibly 100 workers, can turn out tens of thousands of identical, complex chips per day. Although many must be discarded as rejects after testing, the efficiency is so great that a microprocessor capable of directing a car's ignition and fuel systems may cost less than a tank of gasoline.

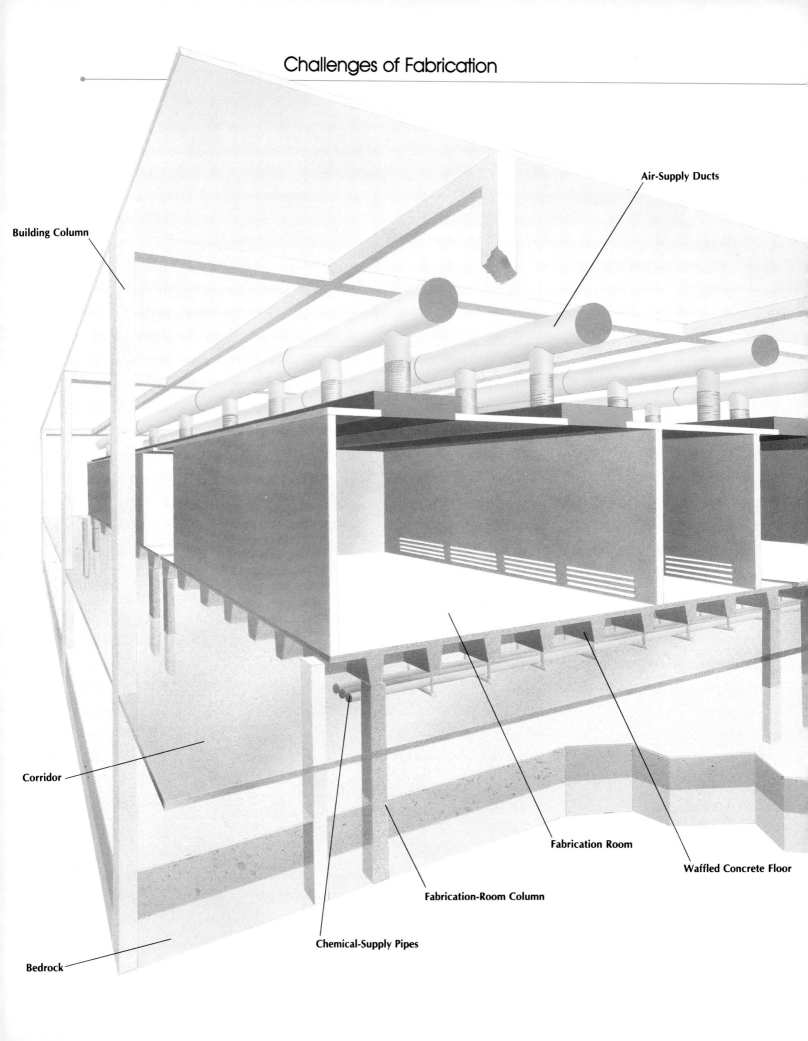

Building Column

Air-Supply Ducts

Corridor

Fabrication Room

Waffled Concrete Floor

Fabrication-Room Column

Chemical-Supply Pipes

Bedrock

A Building within a Building

Like a baby in its mother's womb, the fabrication area of a chipmaking factory is isolated within its own protected environment—sealed off from atmospheric contaminants *(following pages)* and cushioned against damaging jolts *(left)*. But far more than even a baby, fabrication rooms must be elaborately shielded from movement. Chip elements are so small—as narrow as one-hundredth the diameter of a human hair—that just the vibration from a footfall could knock them far out of line during processing.

To block the vibrations generated by people and factory equipment, to say nothing of automobiles and trucks, an unusual structural design has been developed. The fabrication rooms in a typical plant form a building within a building. Each is structurally separate from the others and from the roof, offices, and corridors of the main building that surrounds them all. Machines such as pumps and air conditioners are located as far away as possible, and may even be installed in buildings of their own.

Each room is a boxlike structure on what amounts to a concrete table. The tabletop is heavy concrete, with a tiled upper surface for a floor and a waffled design underneath; the ridges of the waffle block lateral vibration, while the thin areas between ridges simplify drilling for pipes. The tabletop is supported by four legs set into bedrock. Only an earthquake would make itself felt inside.

The unshakable workspace. Rooms for chipmaking rest on two-foot-thick waffled concrete tables *(yellow)* with legs sunk twenty feet into bedrock. All of the surrounding building, including corridors, is separate, supported by its own columns. Ducts *(light blue)* that supply air to the sealed rooms through fine filters *(dark blue)* are hung from the surrounding building and connected to the fabrication rooms with duct sections made of resilient rubber. Similar rubber buffers isolate each tabletop from its neighbors, pipes under the floor, and the surrounding structure, blocking interchange of vibration.

Rooms That Are Cleaner than Clean

Where objects are measured in micrometers, specks of dust are boulders. They can block circuitry in a chip or gouge holes in thin layers during manufacture. To prevent such contamination, fabrication rooms must also be clean rooms. Here cleanliness takes on a wholly new meaning. Even fresh air normally has in every cubic foot some 15 million potentially harmful dust specks, those half a micrometer across or larger. In some clean-room areas, chipmakers raise air cleanliness to "class-one" conditions—no more than one particle 0.3 micrometer or smaller per cubic foot. By this standard, even the air in hospital operating rooms is filthy, typically containing 2,000 particles per cubic foot.

Achieving class-one conditions requires complex measures. The only air that comes into the room enters from the supply ducts after filtering has made it 99.99 percent pure. Separate exhaust systems remove fumes from chemical processing inside the room. There, a more serious source of contamination is human. People not only bring in dust on their bodies and clothing, but themselves are veritable dirt factories, shedding lint, skin flecks, hair, and salt-laden perspiration. To eliminate such particles, workers must be encased in lint-free "bunny suits," face masks, booties, and gloves. Even notepads are made of lint-free synthetic paper.

Not only must foreign materials be kept under control, but so must the chemicals employed in chipmaking processes, since the substances can be life-threatening. Among the chemicals used may be arsine (an arsenic-hydrogen gas), hydrofluoric acid—so corrosive it eats away glass—and an even deadlier mixture of nitric and hydrofluoric acids. Other materials and techniques are similarly dangerous, requiring extensive precautions to prevent harm to workers and the environment around the plant.

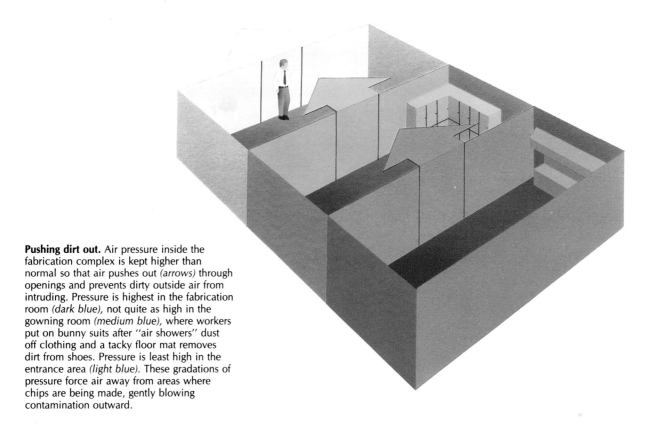

Pushing dirt out. Air pressure inside the fabrication complex is kept higher than normal so that air pushes out *(arrows)* through openings and prevents dirty outside air from intruding. Pressure is highest in the fabrication room *(dark blue)*, not quite as high in the gowning room *(medium blue)*, where workers put on bunny suits after "air showers" dust off clothing and a tacky floor mat removes dirt from shoes. Pressure is least high in the entrance area *(light blue)*. These gradations of pressure force air away from areas where chips are being made, gently blowing contamination outward.

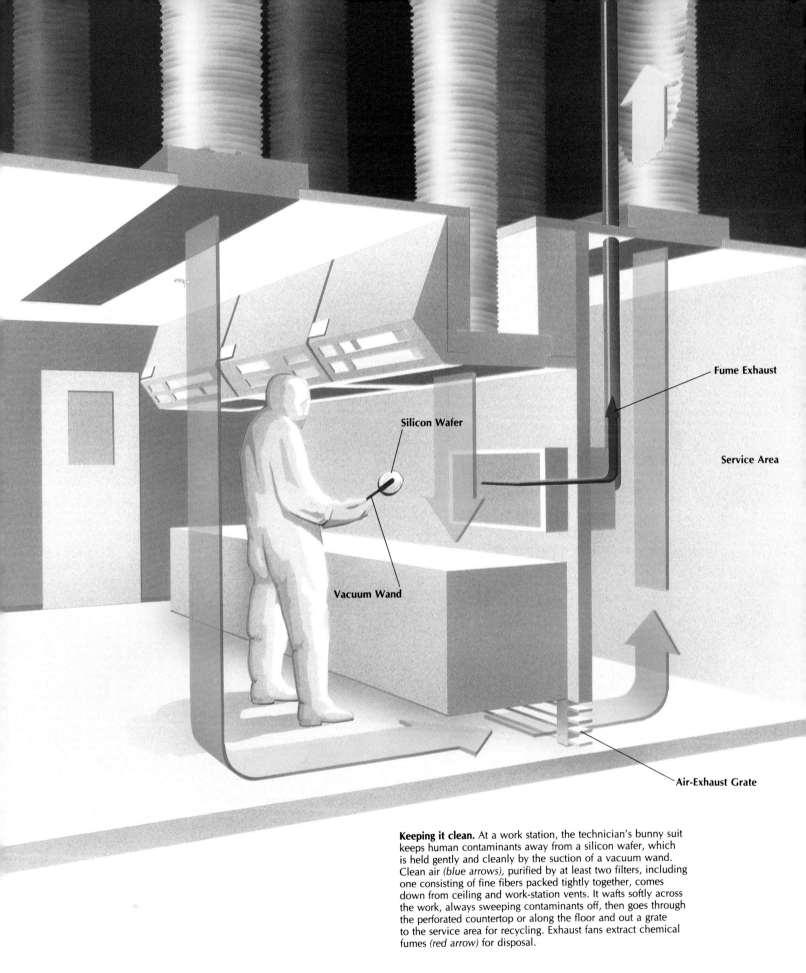

Fume Exhaust

Service Area

Silicon Wafer

Vacuum Wand

Air-Exhaust Grate

Keeping it clean. At a work station, the technician's bunny suit keeps human contaminants away from a silicon wafer, which is held gently and cleanly by the suction of a vacuum wand. Clean air *(blue arrows)*, purified by at least two filters, including one consisting of fine fibers packed tightly together, comes down from ceiling and work-station vents. It wafts softly across the work, always sweeping contaminants off, then goes through the perforated countertop or along the floor and out a grate to the service area for recycling. Exhaust fans extract chemical fumes *(red arrow)* for disposal.

An Electronic Layer Cake

In the calm, pristine conditions of a fabrication room, integrated circuits are cooked up like a layer cake, five to twenty sheets of various micrometer-thin materials placed on a silicon base about one-fiftieth of an inch thick. The base is a wafer several inches in diameter; when the layering process is complete, the wafer will be divided into a multitude of individual—but identical—chips.

Some layers are insulators, some are semiconductors, and others are metallic conductors such as aluminum or gold. They become electronic components such as transistors, resistors, or diodes. Still others are temporary layers *(following pages)*, removed after they have protected areas beneath to define them for the processing that creates circuit components and interconnects all layers.

Equally various are the methods that are employed to produce the layers. Heating can vaporize metals to condense on the chips; electrically charged gas atoms can knock particles off a layering material so that they deposit on the chips, a process called sputtering; or in a method called chemical vapor deposition *(right)*—often preferred because it permits delicate control—gases react to form a vapor that crystallizes on the surface of the chips.

The Layers of a Chip

The schematic drawing at right shows the layers put down successively on a silicon base *(gray)* to make up one small component of a chip, while the drawing at far right depicts the actual component after layering and other treatments are completed. The first two layers *(purple)* aid in the "doping" treatment that creates electronically active areas *(pages 84-85)* in the base, while the third layer is a protective "photoresist" *(lavender)* that defines areas to be etched away or doped *(following pages)*. After treatment, all three layers are removed, as are other photoresists when they have served their purpose. Yellow layers are silicon dioxide insulators; red is polysilicon, a form of silicon used for some circuit elements; blue is metal for conductors. The top photoresist defines insulator areas to be etched for contacts.

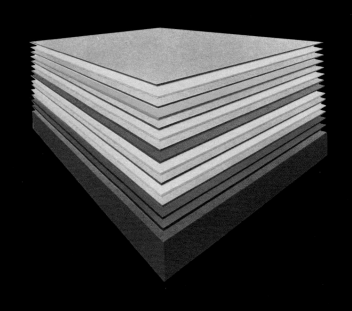

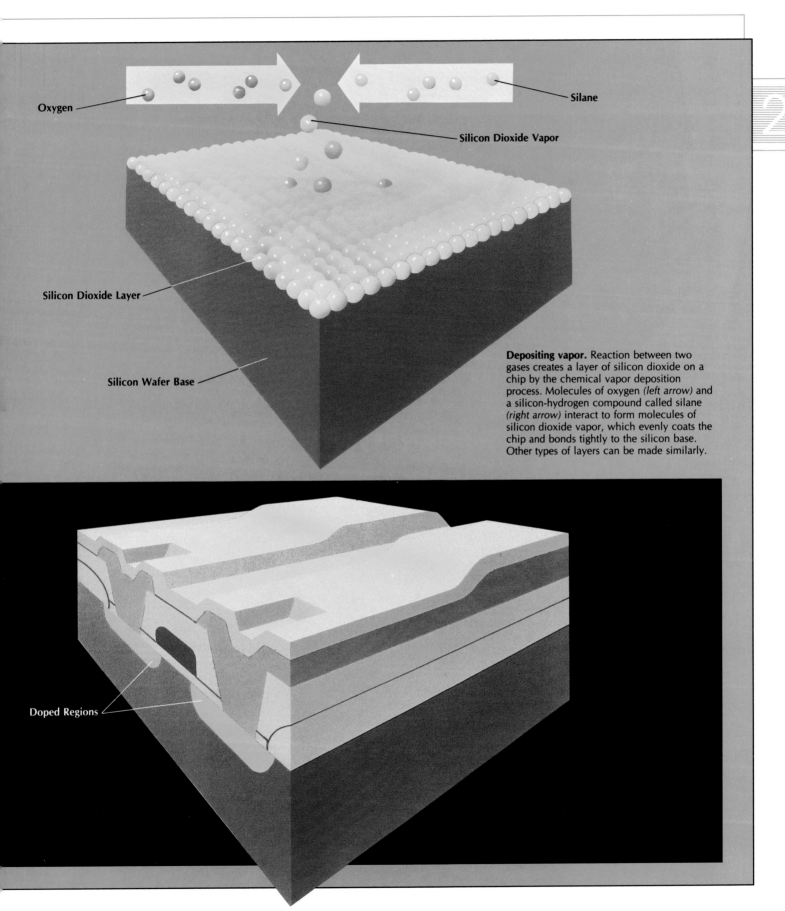

Oxygen

Silane

Silicon Dioxide Vapor

Silicon Dioxide Layer

Silicon Wafer Base

Depositing vapor. Reaction between two gases creates a layer of silicon dioxide on a chip by the chemical vapor deposition process. Molecules of oxygen *(left arrow)* and a silicon-hydrogen compound called silane *(right arrow)* interact to form molecules of silicon dioxide vapor, which evenly coats the chip and bonds tightly to the silicon base. Other types of layers can be made similarly.

Doped Regions

Challenges of Fabrication

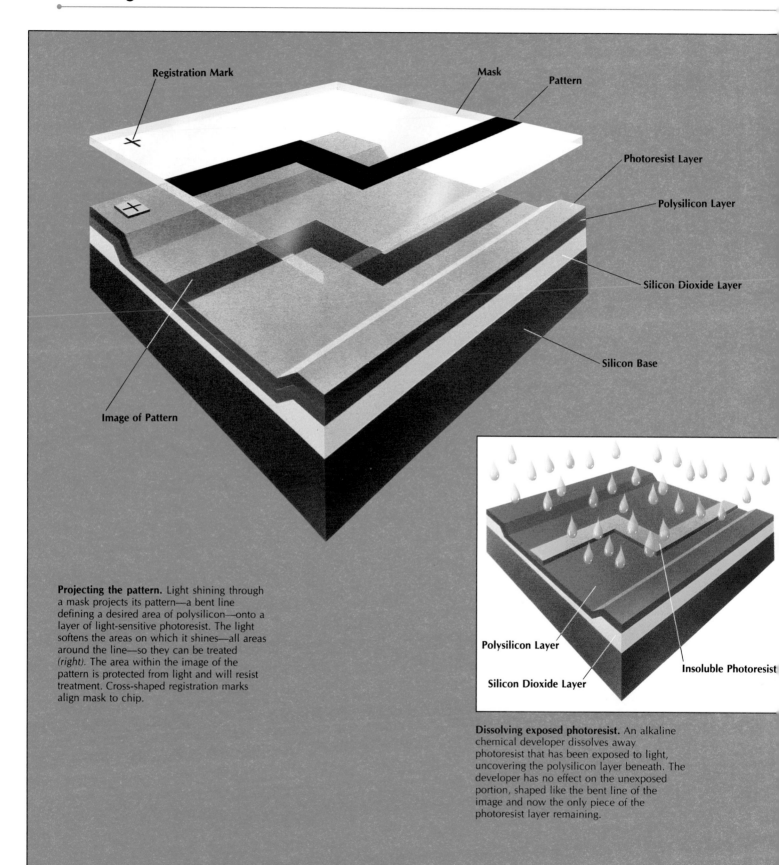

Registration Mark

Mask

Pattern

Photoresist Layer

Polysilicon Layer

Silicon Dioxide Layer

Silicon Base

Image of Pattern

Polysilicon Layer

Silicon Dioxide Layer

Insoluble Photoresist

Projecting the pattern. Light shining through a mask projects its pattern—a bent line defining a desired area of polysilicon—onto a layer of light-sensitive photoresist. The light softens the areas on which it shines—all areas around the line—so they can be treated *(right)*. The area within the image of the pattern is protected from light and will resist treatment. Cross-shaped registration marks align mask to chip.

Dissolving exposed photoresist. An alkaline chemical developer dissolves away photoresist that has been exposed to light, uncovering the polysilicon layer beneath. The developer has no effect on the unexposed portion, shaped like the bent line of the image and now the only piece of the photoresist layer remaining.

Sculpting the Circuit Pattern

To make a chip, an intricate pattern is sculpted into layers of different materials, each one of which serves a different function. Specific regions of some layers must be opened for doping treatment *(following pages)* while other regions are kept closed. Electronically active regions must be connected by metal strips, within a layer and between layers, while other regions must be insulated. This pattern is generated by a masterful refinement of the old printing process of photolithography, illustrated in the sequence beginning at left.

Photolithography in chipmaking works by projecting a pattern on a layer of light-sensitive plastic, the photoresist, for treatments that will open up (or close off) parts of the next layer beneath. Fineness of detail depends on the wavelength of light being shorter than the smallest feature in the pattern; using a longer wavelength would be like trying to separate one grain of sand from a beach with a gloved hand. Visible light ranges in wavelength from about 0.4 to 0.7 micrometer, close to the size of many chip sections. For some sections, ultraviolet light—wavelength about 0.3 to 0.4 micrometer—is employed. And for greatest precision, chipmakers are investigating the use of x-rays with wavelengths between 0.001 and 0.004 micrometer.

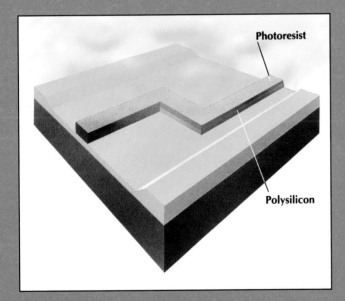

Etching polysilicon. Acid or a gas of charged molecules of a chlorine-fluorine compound *(yellow cloud)* etches away the exposed polysilicon layer where it had been covered by photoresist (outside the pattern), baring the silicon dioxide layer. This etching has no effect on the remaining photoresist or the polysilicon under it, the only remaining piece of the polysilicon layer. The bent line now has two layers, photoresist atop polysilicon.

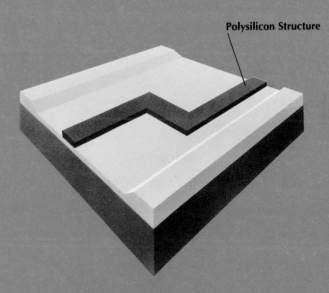

The final step. After the last piece of photoresist is dissolved away by a solution containing hydrochloric acid, the bent-line polysilicon structure remains as a segment running across a layer of silicon dioxide. The sequence for making this segment is finished and the chip is ready for additional operations—doping, layering, and patterning—necessary to finish the circuit element.

Doping to Make the Chip Work

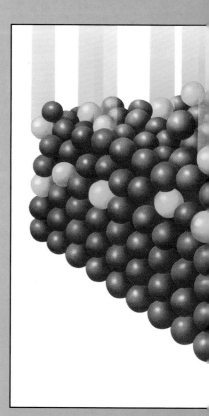

Shooting ions in. During an ion implantation process, electrically charged dopant atoms *(green)* are forced by an electric field into an area of the silicon base *(gray)* where the protective silicon dioxide layer *(yellow)* has been etched away. The remaining silicon dioxide keeps the ions out of the areas beneath it.

Before doping. Atoms in a slice of silicon crystal are in perfect order, each atom *(gray)* occupying a site known as a lattice point.

During the fabrication process, the silicon wafer is deliberately adulterated—doped with impurities in order to adjust its electrical properties. Some dopants, such as phosphorus, give the silicon an excess of electrons, making it n-type, while other dopants, such as boron, create a deficiency of electrons, making the silicon p-type. The two types of dopants, each introduced into the silicon in tiny regions that are precisely defined by the photolithographic sculpting process, produce the circuit components that enable the completed chip to perform its electronic magic.

Doping can be done by diffusion, in which a hot vapor of impurity atoms penetrates the silicon and is driven farther in by more heating. In a newer technique, ion implantation (left), dopant atoms are first stripped of an electron to become positively charged ions, then electrically accelerated in a vacuum so that they force themselves into the silicon. This method places atoms very precisely, and because they are charged, permits exact measurement of the amount added. But it disrupts the silicon crystal's orderly arrangement, which must be restored by annealing (below).

During doping. When dopant ions (green) shoot into the silicon crystal, they strike some of its atoms, knocking them away from their lattice points, wrecking the perfect crystalline arrangement, and creating a disorderly mix of atoms of dopant and silicon.

After annealing. Heating jiggles all the atoms around so that they can fall into their most natural positions—the lattice points. The crystalline order is now restored, but with dopant atoms now occupying some points in the crystal that formerly held silicon atoms.

Probing for flaws. A finished wafer, held on a chuck by suction, is inspected automatically. Probes reach out like tiny fingers from the test instrument that surrounds each chip in turn. Each probe, driven by a computer that feeds signals and records responses, touches a different contact pad on the chip. The rainbow colors are an optical effect created by light refracted by the thin "passivation" layer of silicon dioxide—glass—that coats each chip.

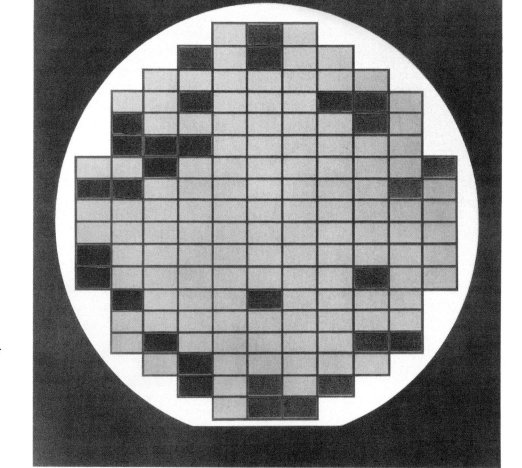

Identifying good and bad. A map, generated by the computer from data gathered by the inspection probes and stored in its memory for later use, shows the location of good (gray) and bad (red) chips on the wafer. In plants that are less fully automated, the inspection machine simply deposits a drop of ink on each reject, which can later be picked out by a human worker using a microscope or by a machine that bonds the chip to a supporting platform (pages 110-111).

Picking Out the Good Chips

After a silicon wafer has been made into dozens or hundreds of completed chips, they are tested and cut apart, and the good ones are selected for packaging *(pages 109-119)*. In some chipmaking plants, this process is almost fully automated, as illustrated here.

Every chip must be examined individually, in some cases by a machine that runs circuits through a sequence of tests at the rate of about 10,000 a second. But the separate parts of a chip—the transistors, resistors, diodes, and so on—are too small to be reached by test probes. These parts are checked at several earlier points in the fabrication process by means of several special chips, having much larger features than the others, that contain components created by the same process that produced the regular ones. Circuit elements on these "test dice," big enough for simple probing, serve as samples to indicate the quality of all parts in the wafer.

Testing is crucial because it feeds back data, such as the location of particular kinds of failure, that enable engineers to improve fabrication processes by altering mask designs or changing dopant concentrations. And yet, despite all this care—computerized controls, elaborate testing, ultraclean and vibration-free workrooms—the rejection rate would shock manufacturers of other products. Defects ruin crystals; dust sometimes gets in; connections are missed. Half of some chip batches must be discarded; a satisfactory yield is 70 to 80 percent usable chips. Only the automated replication of a chip hundreds of thousands of times makes it inexpensive enough to become a ubiquitous part of modern life.

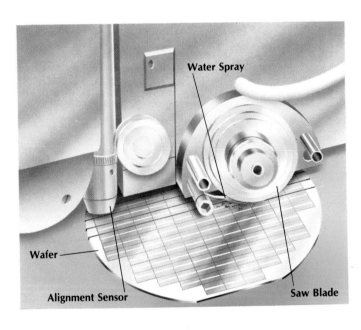

Dicing chips. A diamond-edged saw (depicted here with its cover removed) cuts apart the individual chips, or dice. The machine, guided by an alignment sensor wired to the computer, slices through the wafer along the .005-inch-wide "streets" between chips. Water sprays on the blade and the wafer being cut to cool them and flush away silicon sawdust.

Extracting the good ones. A die picker uses suction to lift only the good chips from the wafer. The machine is guided in its selections by the computer, which uses the wafer map *(far left)* generated by the testing sequence.

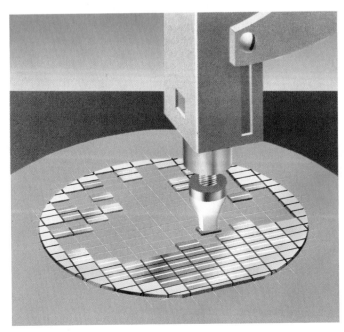

The Quest
for Speed

Customers of the computer industry express many cravings—for lower prices, more capacious storage devices, more versatile software, easier exchange of information among machines, and so on. But nothing is more sought-after than speed. Faster is always better, it seems. A quicker computer not only can do more of everything, but often can take on entirely new jobs. In a word, speed is the computer imperative.

The quest for faster digital processing proceeds on many fronts. Some researchers are exploring materials that let electrons, the messengers inside the computer, travel faster than they do through silicon. Others focus on the design of special-purpose chips that combine a narrowly defined set of functions in a single integrated circuit that replaces dozens—or even hundreds—of general-purpose ICs wired together to perform the same work. And speed is the push behind continued efforts toward ever-denser chips. No sooner had manufacturers unveiled a chip containing a million transistors—the so-called megachip—than they and their eager customers began a countdown to the advent of the gigachip, a prospective integrated circuit comprising a billion elements. Someday in the not-too-distant future, their vigil will be rewarded with a supercomputer on a chip, containing the microprocessor, input/output ports and controllers, special processors for mathematics and high-resolution color graphics, megabytes of memory, and all the other ingredients that now fill a sizable cabinet. Building these capabilities into a single chip might multiply computer speed twentyfold, simply by reducing the average distance that electrons must travel as they weave their tapestry of logic.

Major obstacles confront gigascale integration. Designing such a chip, for example, is a task that computer scientists would call "nontrivial." Indeed, without substantial assistance from computers, laying out this chip might never be economically justifiable, regardless of how much faster it promises to be. Even if the difficulties of designing a billion-component chip were to evaporate overnight, actually making it would pose prodigious challenges. To gain maximum benefit from gigascale integration, the billion elements need to fit onto a wafer of silicon little larger than the ones presently used for million-component ICs. In megachips, the width of a conductor or other chip feature may be as little as one micrometer. For the gigachip, this measurement must be quartered, a feat requiring fabrication technology and equipment that may not be ready until near the turn of the century.

Finally, there is the question of reliability: Features so small may exceed the limits of the materials from which they are made. For example, the narrower an aluminum conductor on the surface of a chip, the more difficult is the job of making certain that it contains no gaps to break the circuit. At the incredibly miniaturized scale envisioned for gigachips, the problem develops a new wrinkle: Even if the conductors emerge from the chip factory without breaches, they may not remain continuous when the chip is put to work. Electron pulses may drag atoms of aluminum along with them, creating voids where there were none and causing the chip to fail.

Less power might be the answer; a smaller current would be less likely to disturb aluminum atoms. However, there is another rub. The atoms in the conductors and elsewhere in a chip, constantly vibrating because of heat generated inside a computer, give rise to spontaneous, random currents. The pulses sent intentionally through the computer must be stronger than this electronic noise, or the computer will mistake unintentional pulses for data. Such confusion would be disastrous.

SUPER SUPERCONDUCTORS

If a computer could be built of components that did not become hot in operation, electronic noise might be kept to a minimum, and pulses could be weaker, yet remain intelligible. Components might then be made finer and placed even closer together, taking the chip a step toward gigascale integration. Since heat arises in a computer from electrical resistance within its electronics, the machine would run cooler if resistance were reduced. Were it to be entirely eliminated, heat, like the vacuum tube, would become a computer relic. Such is the attraction of superconductivity, the ability of some substances to offer no resistance whatever to an electric current.

Superconductors have tantalized scientists and engineers ever since Heike Kamerlingh Onnes, a Dutch physicist, first discovered the phenomenon in 1911, when he cooled mercury to within three degrees of absolute zero, the Kelvin-scale temperature at which the incessant vibration and jostling of molecules against one another ceases. (On the Celsius scale, absolute zero is 273 degrees below the freezing point of water.) To his amazement, Onnes found that the ultracold mercury carried electricity as though it were passing through a vacuum. Electrons traveled as fast as light fleets through space—four times what they could manage in a piece of copper wire. (Silicon is a far slower conductive medium. Passing through a silicon transistor, electrons poke along at one three-thousandth of light's speed—just sixty-two miles per second.)

A superconductor behaves like any other conductor until it is cooled to a point called its critical temperature. Then the resistance abruptly drops to zero; current will flow through a circuit indefinitely, with no loss of energy. Resistance remains at zero as long as the temperature and two other critical factors are not exceeded. If a superconductor is exposed to a magnetic field stronger than a critical limit, resistance reappears. A similar limit applies to the amount of current traveling through the material. If the flow is too strong, superconductivity disappears.

Although Onnes predicted that mercury "films of molecular dimensions" would carry large currents, or that wire coils of other, as-yet-undiscovered superconductive materials would generate huge magnetic fields useful in electric motors, generators, and scores of other tools of the industrial age, he and those who followed made no headway against two formidable barriers to technological revolution: superconductivity appeared only near absolute zero; and superconductors could only carry currents that were so small as to be useless. Even if the current could be increased, keeping a superconductor cold was a costly business that involved the use of liquid helium. Helium gas is itself a rare and costly commodity, constituting only one part in 200,000 of the atmosphere; to reduce it to superconductor temperatures requires a process that involves first compressing and cooling hydrogen, then using it to liquefy compressed helium. Only when this liquid is allowed to evaporate and expand does it achieve its uniquely low temperature.

For these reasons, superconduction remained a scientific curiosity until the 1960s, when Bell Laboratories researcher Gene Kenzler demonstrated that an alloy of niobium and tin, when cooled to a relatively tepid twenty degrees Kelvin, could carry as large a current as a copper wire of the same diameter could conduct at room temperature. Scientists looked upon the twenty-degree Kelvin critical temperature of the alloy as encouraging, though use of the material would be expensive. They soon found ways to put the new superconductor to work: Niobium-tin superconductors are used to power giant electromagnets essential to experiments in the control of nuclear fusion reactions, and to guide charged atomic nuclei as they zip around particle accelerators. Superconductive magnets also lie at the heart of magnetic-resonance imaging equipment that gives doctors accurate three-dimensional images of organs in the human body. This machine is one of the most common and perhaps least expensive superconductor devices to operate, yet the annual bill for the liquid helium and liquid nitrogen required to operate a magnetic-resonance imager is $60,000.

For computers, the most promising development was a 1962 invention by Brian Josephson, a Cambridge University graduate student who had attended a series of lectures by Bell Laboratories' Philip Anderson, who presented his theories of electron movement. Josephson combined Anderson's lectures and his own study of quantum mechanics to come up with a superconductive switch called a Josephson junction. Eleven years later, Josephson would win the Nobel Prize for his work on the device.

The Josephson junction consists of two superconductors separated by a thin insulating layer. As long as a magnetic field is present around the device, superconduction is halted and current cannot pass through the insulator. Interrupting the magnetic field allows the electrons to pass through the insulator, turning the switch on.

This passage of electrons takes place because of the unique way they pair up in a superconductor. Not only are crystals of superconductors aligned in such a way that they offer very little interference to electrons, but according to the most prominent theory of how superconductivity occurs, the electrons themselves are linked in pairs that speed each other along. If one gets stuck, the other pulls it forward. The electron pairs, being fragile, dissociate when vibrated by even a small amount of heat or when placed in a magnetic field. But while they exist, they are able to tunnel through the insulator, popping across the insulating barrier without help from any outside force, such as the voltage that must be applied to an ordinary circuit. Josephson's switches are 100 times faster than transistors in their ability to turn on and off.

Speed being an essence of computing, the Josephson junction attracted the attention of IBM, Bell Laboratories, and Sperry Univac in the early 1970s. IBM researchers pursued a practical application of the Josephson junction to computers for more than a decade, but their attempts failed to overcome the persistent problems of temperature, current capacity, and cost that seemed to be inherent in superconductors. In addition, they encountered a serious obstacle, unique to

Josephson's junction, which compounded the difficulties of employing it in computers. It can act only as a switch, whereas a transistor can be used not only as a switch, but as an amplifier, a diode, a capacitor, or a resistor. Although the ability of Josephson's invention to measure very small magnetic fields made it valuable for a variety of test and research instruments, the switch remained a switch, and nothing more. After spending perhaps as much as $300 million on the project, IBM abandoned the Josephson junction in 1983, with the terse announcement from John Armstrong, an IBM vice president for logic and memory, that superconductivity was "not a winner technology." Sperry and Bell Laboratories followed suit.

A WARMER REFRIGERANT

The cost and difficulty of maintaining extremely low temperatures remained the principal obstacle to the use of superconductors in computers; even without Josephson junctions, higher-temperature superconduction could do much to speed signals between a chip's components or between chips. But the gap between reality and practicality was considerable. A really useful superconductive material would need to have a critical temperature no lower than seventy-seven degrees Kelvin, the temperature of liquid nitrogen. Nitrogen is plentiful, cheap, and easy to handle. Scientists routinely store it in ordinary vacuum bottles.

The next step toward a seventy-seven-degree superconductor came, ironically, from an IBM researcher, Karl Alex Müller. Working at a laboratory in Zurich, Switzerland, Müller held the title of IBM Fellow, a much-coveted corporate honor that freed him to pursue projects of his own. His interest was a class of ceramics called perovskites, and he wanted to look into the strange electrical behavior of some that he had seen. These compounds, normally insulators, had sometimes, unpredictably, behaved like metal conductors. For three years, he and colleague Johannes Georg Bednorz quietly and laboriously mixed and tested hundreds of compounds for origins of superconductivity. "We didn't tell anybody what we were doing," said Bednorz, "because we would have had difficulty convincing our more knowledgeable colleagues that what we were doing wasn't crazy."

In January 1986, Müller and Bednorz hit upon a substance that earned them the 1987 Nobel Prize in physics—a superconductive ceramic compound made of barium, lanthanum, and copper oxide. It had a critical temperature of thirty degrees Kelvin—ten degrees higher than niobium-tin. It was a stunning leap forward, and scientists all around the world immediately sought further gains. Within two years there were dozens of new claims of compounds with critical temperatures exceeding the seventy-seven-degree barrier. In May 1987, a

Karl Alex Müller (left) and Johannes Georg Bednorz, shown in their Zurich lab, achieved a scientific breakthrough with their 1986 discovery of a superconductor able to function at minus 238 degrees Celsius, at that time the highest operating temperature known for such materials. Superconductors that function at even higher temperatures may lead to significantly swifter, more densely packed integrated circuits.

team headed by Paul Chu of the University of Houston reported mixing a ceramic compound that had a critical temperature of 225 degrees Kelvin, as warm as a very cold day in Minnesota.

Such results hold huge technological promise, but they have not yet yielded commercially useful products. The experiments have been difficult to reproduce, and the materials themselves tend to be extremely unstable. Researchers complain of materials inexplicably losing superconductive properties overnight, or watching laboriously produced ceramics dissolve in spilled water. Some of the materials are brittle or crack easily, making them exceedingly difficult to fabricate into wires and other useful shapes.

Yet the quest continues. Researchers at Bell Laboratories have formed flexible wires and tapes from a ceramic superconductor, and IBM technicians have succeeded in growing thin films of a superconducting crystal—an important step toward incorporating superconductors into chips. Scientists at Sandia National Laboratories in Albuquerque, New Mexico, have concocted a nitrogen-cooled superconductor that can withstand a current strong enough for electronic applications. Given such advances, there is good reason to expect that superconductor chips will eventually find their way into computers, especially those demanding the very highest speeds possible.

A FLASH OF INSIGHT

Superconducting alloys are not the only materials that hold the promise of more computer speed. Big payoffs in digital-processing power are already coming from a semiconductor that for thirty years played second fiddle to silicon. In 1985, two years after IBM's Armstrong had dismissed superconductors, he announced that a new material had taken over as "the largest alternative technology" in the computer maker's research laboratories. IBM and a score of other companies had rediscovered gallium arsenide.

That gallium arsenide might have the properties of a semiconductor was the brainstorm of Heinrich Welker, head of a research team at the giant German electronics firm, Siemens, in the 1950s. Well aware that the semiconductors silicon and germanium—as well as other elements such as carbon, tin, and lead, all of which behave as semiconductors to some degree—lie in the same column of the periodic table of elements, Welker wondered whether chemical compounds made by mixing elements from the columns on either side of the semiconductor column might not produce a new family of semiconductors. He chose to combine gallium and arsenic, which stand on either side of germanium in the periodic table. Specifically, he expected that gallium arsenide would prove to be a semiconductor like germanium.

And so it is, with an important difference: Gallium arsenide electrons are more mobile than those of germanium or silicon. One consequence is that electrons travel through gallium arsenide six times faster than they do through other semiconductors. A chip made with gallium-arsenide transistors could be expected to operate as much as six times as fast as a silicon chip. For many years, however, computer builders were unable to take advantage of gallium arsenide's superior speed because the compound eluded efforts to make large quantities in a form pure enough to function as a semiconductor.

Gallium arsenide's problems as a practical semiconductor are rooted in its

Master Plans for Wiring a Computer

A computer's speed depends in part on the time required for data to be shuttled from point to point, not only within a chip, but also between the many chips that populate a circuit board. To maximize speed, chip and circuit-board designers seek to keep the paths as short as possible. This complicated task is done with the help of computers.

Programmed with a few rules, a computer repeats a procedure that identifies efficient routes between points. One universal rule is that wires cannot touch. For simplicity in the examples shown here, another rule is that routes are to be established by moving from square to square across a grid established on the surface of a chip or circuit board. In actuality, additional pathfinding options, such as diagonal routes or the ability of a wire to burrow under an obstacle instead of detouring around it, are common *(pages 114-117)*.

The computer follows an elementary procedure. Beginning at the grid square above the one occupied by a starting point—and working in a clockwise direction—the computer assigns numbers to the surrounding squares. From the resulting quiltlike pattern pops the shortest path *(below)*.

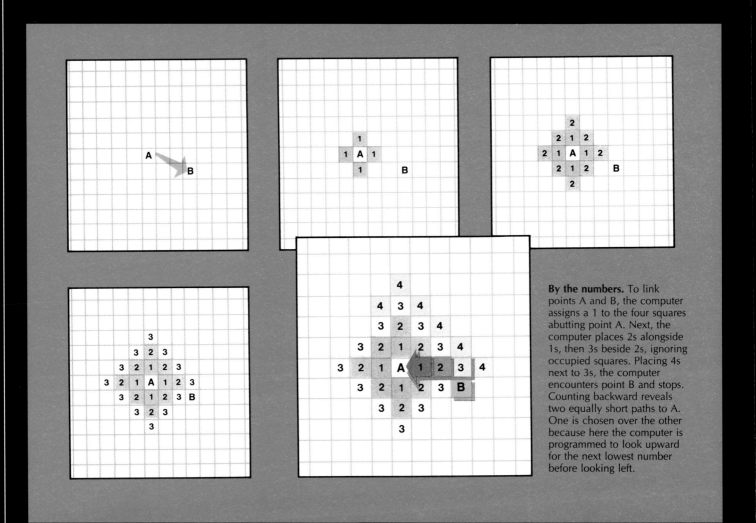

By the numbers. To link points A and B, the computer assigns a 1 to the four squares abutting point A. Next, the computer places 2s alongside 1s, then 3s beside 2s, ignoring occupied squares. Placing 4s next to 3s, the computer encounters point B and stops. Counting backward reveals two equally short paths to A. One is chosen over the other because here the computer is programmed to look upward for the next lowest number before looking left.

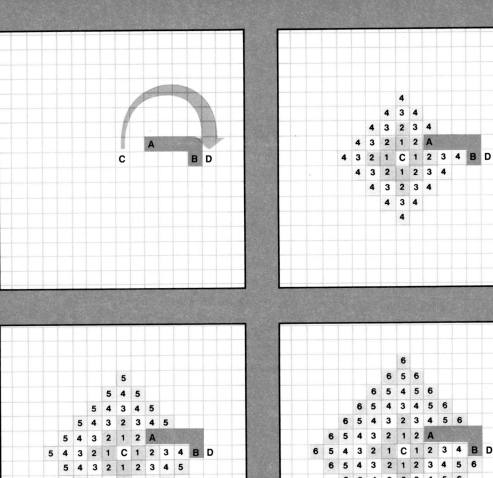

Making a detour. The numbering process shown on the opposite page also finds the shortest way around an obstacle—in this instance the path established between A and B, which blocks the most direct route between C and D. Beginning at C, the computer again assigns numbers to any empty squares it finds. As the pattern of numbered squares expands, it engulfs the path between A and B, encountering D in the eighth repetition of the process. Counting backward identifies a path that makes an end run around B.

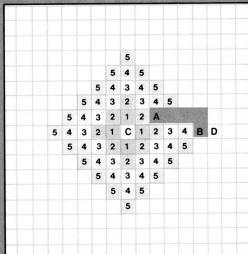

An Embarrassment of Riches

The number of possible solutions for making connections on a chip or a circuit board can become astronomical. Deciding upon which solution is the best is the sort of number-crunching computers excel at. With the most complicated wiring schemes, the work is often done at night, when a computer at a chip-design facility would otherwise be idle. Laboring into the wee hours of the morning, the computer endeavors to concoct the best possible wiring plan.

For a while, the computer may be instructed to proceed according to one set of rules, then to switch to another. It might, for example, start out using the so-called greedy algorithm, which instructs the computer to begin by connecting the points farthest apart before proceeding to link those that are closer together. Later, the computer may reverse its approach and first try joining points that are near each other.

As the work progresses, the computer reroutes some wires to make paths for others. The result is a single wiring scheme. The kind of judgments that the computer makes are illustrated on these pages, with the best solution appearing at right, at the bottom of the page.

In some instances, the computer cannot make all the necessary connections, usually because not enough space has been set aside for the wiring necessary to link components in a design. At this point humans intervene, searching for wiring opportunities that the computer might have overlooked. If they cannot be found, then the design must be revised and the computer asked to find a solution to the new problem.

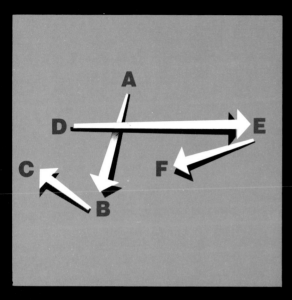

A range of choices. Joining the six points in alphabetical order—A to B to C and D to E to F—is one solution to the problem, but this approach yields a less-than-optimal result *(below)* compared to the layouts that result if the sequence is left to the computer *(right)*.

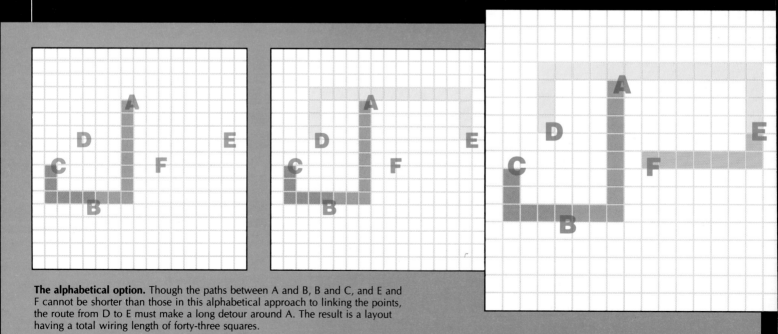

The alphabetical option. Though the paths between A and B, B and C, and E and F cannot be shorter than those in this alphabetical approach to linking the points, the route from D to E must make a long detour around A. The result is a layout having a total wiring length of forty-three squares.

An improved sequence. By connecting D to F before proceeding to E, the computer saves the distance that the alphabetical option wastes in doubling back to reach F from E. Total length: thirty-nine squares.

The value of a branch. For this variation of the solution at left, the computer has been permitted to send an offshoot toward E before reaching F. Electrically equivalent to the preceding effort, this approach is only modestly superior. Total length: thirty-seven squares.

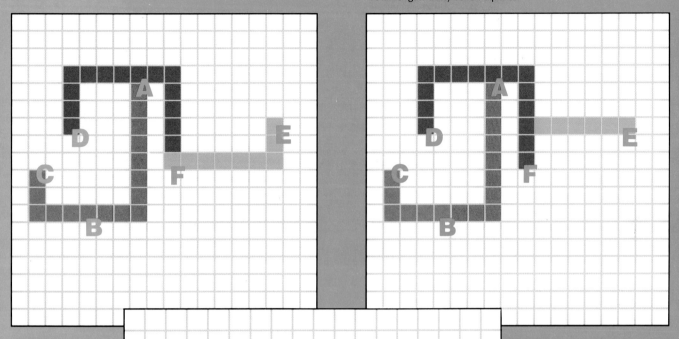

Starting in the middle. The best outcome is achieved when the computer begins by joining D directly to E. A short branch reaches down to F, and the path joining A, B, and C is one square longer than it was in any of the preceding solutions. Total length: thirty-two squares.

nature as a compound of two elements. Gallium is a soft, metallic by-product of aluminum smelting. Arsenic is a highly toxic leftover from copper refining. Among their other differences, the two elements boil at different temperatures, and so, when the compound is melted to make a single crystal, the arsenic, having the lower boiling point, starts to evaporate out of the liquid. In the language of the chemist, the elements of the compound dissociate. Small amounts, enough to experiment with, could be made in a laboratory, but the attractiveness of gallium arsenide's speed did not compensate for the difficulty of manufacturing it in the quantities required to make millions of chips.

A COMMUNICATIONS BOTTLENECK

High-speed digital communications prompted the chipmaking world to take another look at gallium arsenide in the 1970s. By that time it had become clear that the microwave channels then available would soon be operating at maximum capacity. Silicon stood in the way of increasing the volume of microwave traffic because the semiconductor limited the speed at which information could be transmitted. Silicon chips could handle a maximum of about two billion bits of data per second. By contrast, gallium arsenide had the potential to raise chip speed to 30 billion bits—ample incentive to reconsider the difficult material.

The purity problem was solved with ingenuity, persistence, and chemical sleight of hand. To prevent the arsenic from boiling away at the high temperature needed to grow a crystal—more than 1,000 degrees Celsius—engineers devised a way to cover the molten gallium arsenide with an atmosphere saturated with an arsenic-rich gas under pressure. With no room in the atmosphere for additional arsenic molecules, the chemical remained in the melt with the gallium.

This solution led to another problem. High temperature and pressure freed silicon atoms from the quartz container used as a melting pot. The silicon atoms constituted an impurity that prevented the gallium arsenide from performing as a semiconductor. Finally, after much trial and error, gallium arsenide manufacturers learned to limit the amount of silicon entering the gallium arsenide by carefully monitoring temperatures. Scientists also found, almost paradoxically, that they could neutralize the silicon by adding small amounts of other impurities, such as chromium.

Having a homogeneous crystal that could be sliced into wafers was only the first step toward a gallium arsenide chip. Next, a new variety of transistor had to be designed. One reason that silicon is commonplace in chipmaking is that it oxidizes easily. A fabricator need only heat silicon with steam to grow an insulating layer of silicon dioxide over its surface. This thin insulator is used to separate a voltage-carrying electrode from the transistor's surface, forming the gate of a field-effect transistor (FET), a vital element in large-scale integrated circuits. But such thin insulating films cannot be grown on gallium arsenide. Chipmakers found, however, that a gate can be formed without the insulating film when they use a metal electrode, rather than the usual polysilicon conductor. The junction of the metal and semiconductor creates an insulating barrier, a property first discovered in 1939 by German physicist Walter Schottky and rediscovered by William Shockley's team as they developed the transistor.

Other snags arose in the fabrication process. Silicon is doped by ion implantation (pages 84-85), a process that involves shooting high-speed charged particles at the chip. This disrupts the crystal structure at the surface, but the damage to silicon is repaired by heating, or annealing, the wafer so the crystals can re-form. However, when gallium arsenide is heated after ion implantation, the arsenic once again evaporates, and the very material of the chip is ruined. To prevent such destruction, gallium arsenide chips must be annealed in a pressurized atmosphere of gallium and arsenic.

But like the silicon chip in 1961, the gallium arsenide chip needs a push down the learning curve if it is to replace silicon or even challenge its preeminence. In the 1960s, it was the space race that fueled development of silicon; the demand for thousands, then millions of miniature, lightweight circuits needed to build and guide the missiles of research and defense gave chipmakers the money and experience they needed to make the integrated circuit a common household item. For gallium arsenide, this push could come from the communications industry's unslaked thirst for higher speed. Gallium arsenide chips have indeed begun to ease the communications crunch. Not only are they widely used in the latest armed forces communications systems, but they are appearing in civilian applications as well—including home satellite antennas for receiving television broadcasts. A hint of what could lie ahead for gallium arsenide chips may be found in the work of Seymour Cray, whose supercomputers set the standard for speed. In 1987, Cray announced that the next generation of computers built by his company would be based on this speedy alternative to silicon-based integrated circuits.

NO RESPITE FOR DESIGNERS
Since the start of their work in the 1950s, chipmakers have found one certain route to increased speed—the packing of ever-greater populations of transistors onto chips and the consequent shrinking of the distances that electrons must travel. While new materials may reduce the significance of these distances as a deterrent to performance, chip designers persist in their efforts toward greater integration. Making chip details as small as possible pays off not just in speed but also in manufacturing economies and greater yields.

The gigachip lies on the far horizon of this effort. A billion of anything is a mind-boggling concept. Connecting such a number of transistors together in a logic network that functions as an entire computer is an impractically complex and time-consuming task, unless chip designers receive substantial assistance from computers. In this respect, important trailblazing has occurred in the realm of relatively simple chips.

In 1982, three young engineers left Commodore International Ltd., maker of the Commodore home computer, to start a competing company. Al Charpentier, Bruce Crockett, and Bob Yannes were already industry veterans, having worked together on the design of two of history's most successful home computers, the Commodore 64 and the Vic-20. However, their plans were dashed within months of their departure from Commodore, when the home computer industry tottered. The result was their entry into an entirely different market—and leadership in an emerging new segment of the chipmaking business.

Yannes suggested that the partners focus on electronic keyboards, a computerized musical instrument with a piano keyboard capable of producing a wide variety of synthesized sounds. An amateur musician, he was unhappy with the high prices of electronic keyboards then available. With better design, he said, the three collaborators could manufacture keyboards that would outperform and undersell the Japanese competition. Ensoniq, Inc. was born.

The keyboard they chose to build was a type known as a sampler—essentially a computer designed to manipulate sounds, rather than words or numbers. It can record a sound, analyze and remember its characteristics, then play it back as "music." A Chopin piano concerto could become, if a musician were so inclined, a concerto for screeching brakes. The notes would be Chopin's, the same as those played on a piano, but the sound would be gratingly different.

Before Ensoniq arrived in the market, the least expensive samplers cost more than $8,000. The new Ensoniq Mirage, introduced in 1985, carried a price tag of $1,695. It sold rapidly, and the next model did even better. The company captured and held a significant part of the keyboard market from such Japanese giants as Casio, Korg, Roland, and Yamaha. Ensoniq had found a way to provide the musician with a sampler that performed at least as well as the competition at one-fourth the cost.

Their secret weapon was a single integrated circuit that took the place of more than a hundred chips and other separate electronic parts used by Ensoniq's competitors. Those companies had followed the usual practice of designing their electronic circuits around standard, multipurpose components. They were relatively cheap and widely available, but assembling a large number of them to

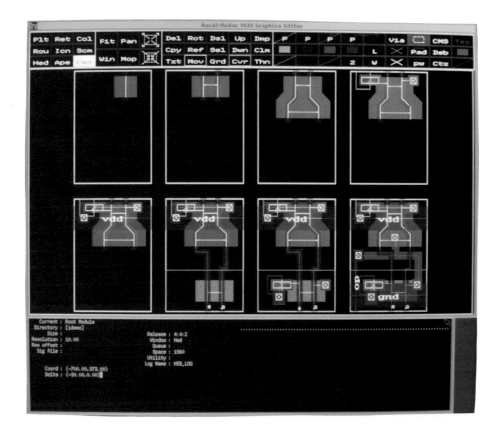

Eight steps in designing a logic gate appear on the screen of a computer as an increasingly complex maze of colors. Light blue and green areas represent doped regions of transistors; colored lines delineate conducting channels. The completed design, known as a standard cell, can be stored by the computer for later use.

make a keyboard was a slow, expensive process. Charpentier designed for Ensoniq an integrated circuit that performed exactly—and only—the functions executed by the scores of parts built into the average sampler. The cost of the chip was about ten dollars. Assembling the Mirage keyboard took seconds. Besides simplifying assembly of the product, the Mirage chip was very fast, accomplishing its signal-processing tasks in about half the time required by keyboards manufactured from general-purpose components, largely because electronic pulses are not forced to travel long distances between chips.

Ensoniq was a pioneer in the design of application-specific integrated circuits, or ASICs. Many other companies have pursued the same strategy. For example, when Raster Technologies, a Massachusetts builder of graphics systems, spent $100,000 in 1984 on the design of a custom chip to replace 250 individual components in one of its products, it saved $700 on each unit. Raster founder Louis Doctor said the company paid for the design with less than one month's production. Other computer-graphics companies have found that specialized chips enable them to build in features unavailable in any other way. Even more promising for the future is the success of engineers at GTE Laboratories in Waltham, Massachusetts, in condensing an entire business switchboard—called a PABX, or private automatic branch exchange—into a single half-inch-square chip containing 350,000 transistors.

ASICs free manufacturers of electronic products from dependence on off-the-

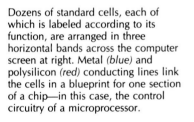

Dozens of standard cells, each of which is labeled according to its function, are arranged in three horizontal bands across the computer screen at right. Metal *(blue)* and polysilicon *(red)* conducting lines link the cells in a blueprint for one section of a chip—in this case, the control circuitry of a microprocessor.

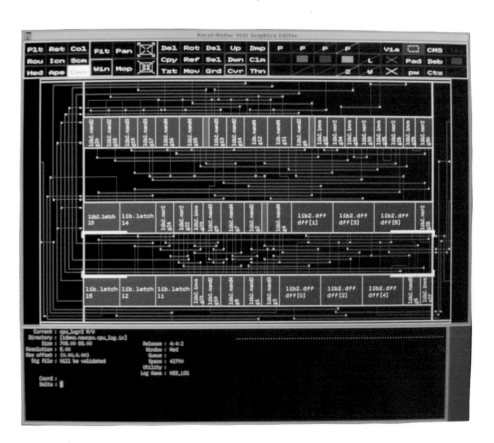

shelf chips produced by the semiconductor giants. Instead, members of a product team design a chip to do exactly what they need, then send it to a silicon foundry for fabrication. The result, says Charpentier, is that "the power of silicon is becoming more usable and readily available to a larger population of small companies." In 1985, about 6,000 ASIC designs were produced; annual output of new designs during the 1990s is expected to be 90,000.

This flood of ASICs would be impossible without low-cost, computerized chip-design systems called silicon compilers *(page 33)*, which can turn almost any electronics engineer into a competent integrated-circuit designer, capable of inventing a chip to meet a special need. Silicon compiler systems allow designers to draw on databases of standardized components, which are automatically assembled into a whole according to mathematical rules. Once extremely limited, compilers are becoming increasingly versatile as computer technology permits the use of larger databases and experience permits designers to refine their assembly rules.

Nothing similar yet exists to aid the engineers charged with designing the highly integrated chips of a million or more—and ultimately a billion—components. While it takes mere hours to concoct a simple ASIC with a silicon compiler, hundreds of designer-years are expended on creating the most complicated integrated circuits, when the goal is the smallest, fastest, and least power-hungry chip possible. To make such high-performance devices economical, especially for applications where demand for the chips is likely to be small, the cost of designing them must be substantially reduced. Computer scientists believe that the means for doing so is to further automate the chip-design process. They hope to bring a measure of human judgment to the undertaking, in part through the use of expert systems—principles of chip design gleaned from the masters and stored in databases as rules for a computer to follow.

A good deal of progress has been made in this direction. At AT&T Bell Laboratories, for example, Dr. Thaddeus Kowalski has headed a team of chip-design specialists working since 1980 on a project that has been named Algorithms to Silicon. The idea, says Kowalski, is "to free the designer to solve the important problems in chip design"—that is, to apply imagination and ingenuity to the problem of defining how a chip is to function and what it is to achieve. More mundane work is the province of the computer. It would decide on the optimal number of registers, adders, and arithmetic logic units (ALUs)—chip components that perform true-false comparisons, among other tasks—and it would handle the tedious minutiae of how to lay out the circuitry.

The shortcoming of silicon compilers has been that, in bringing a designer's description of an integrated circuit to life, they proceeded blindly, step by step. If a design calls for several adders, for example, most silicon compilers dutifully create an adder wherever one is specified—without ever considering whether to combine some of the adders, thus saving space on the chip and, in all likelihood, increasing its speed. In positioning adders and other modular components and in laying out the wiring to connect them, the compiler, proceeding without an appreciation of the entire design, can run into a problem fitting the last module on a chip. To correct the problem, the system has to backtrack to an earlier point and try again—with no assurance that the next attempt will be any more successful than the preceding one.

This hit-or-miss approach is extremely costly in computer time and often results in a second-rate design, a situation, notes Kowalski, "reminiscent of the early chess-playing programs: while it is easy to get a computer to play a legal game of chess, it is exceedingly difficult to get it to play a good game." By using expert systems to complement a step-by-step method, Kowalski hopes to make chip-design software play a better game of chess.

IT'S ALL IN THE DETAILS

The starting point for Kowalski's approach is a complete description of the chip. In the design of a microprocessor, the description would contain details of how the chip will address the computer's memory and the sequence in which the chip will perform the several steps necessary to add a pair of numbers. The specification also includes the maximum dimensions of the chip, how much power it can consume, and even processing delays that may be required to coordinate various chip functions.

In the Algorithms to Silicon scheme, chip specification takes the form of a detailed computer program that is fed into the first stage of the system, a program called Bottom-Up Design. Nicknamed BUD, this software examines all the particulars of the chip description and typically generates several designs that satisfy the requirements. Each one names the semiconductor technology—MOS or CMOS, for example. Each lists, by type, the registers, adders, and ALUs that constitute the design and distributes the chip's various tasks among them. And finally, BUD provides a floor plan for the chip showing how the many components fit together on a rectangle of silicon.

To a degree, BUD can optimize the designs for peak performance. For example, the program might see that a significant increase in processing speed would result from something as simple as positioning registers so that the one whose contents would generally be processed first lay closer to the processor. From the several designs, BUD selects the one that is the smallest, fastest, and least power hungry for the functions that will be used most often, as revealed by a statistical analysis that BUD performs on the chip design's inputs, outputs, and circuitry. The winning design is passed to the system's expert system, the Design Automation Assistant (DAA). Examining the entire specification, the DAA checks it in detail against a database that contains hundreds of rules for the most effective way to fulfill the designer's requirements. Obtained from master chip designers, these rules can be readily changed as human experts refine their skills, making it easy to keep the system up-to-date. The DAA produces a block diagram of the chip in which the various modules appear as labeled boxes connected by lines.

The logic of the chip having been established by BUD and the DAA, the design is processed by a program called a module binder, which fills each of the blocks in the DAA's diagram with predesigned circuits, or cells, extracted from a library of modules based on the various semiconductor technologies. The module binder selects cells that best satisfy the designer's wishes. Should the library lack a cell for one purpose or another, the module binder can combine cells to build the necessary circuitry. A similar program called the control allocator, using other cells in the same library, creates the controller, the part of the chip that regulates the flow of data through the chip.

As detailed as the design has become, it is as yet no more than symbols, the computer equivalent of a circuit diagram drawn on a piece of paper. The final step of the automated chip-design process is to convert this blueprint into actual transistors of silicon, all wired together in such a way that they function as intended. That the design process is able to be be automated, notes Kowalski, "has profound implications. Automatically produced designs can be guaranteed to be correct and can be produced far more quickly than if human intervention were required." Designing a gigascale chip could well become a manageable exercise.

GIGASCALE FABRICATION

Dr. James D. Meindl agrees. Provost at Rensselaer Polytechnic Institute and former head of Stanford University's Center for Integrated Systems, Meindl is often asked to predict the progress of chip integration. Each time, his conclusion is the same: the one-billion-transistor chip is not only possible, it is likely by the end of the century. Meindl's forecast is based on the belief that several prerequisites will be met—that research will yield practical new materials and that designers will develop the automated tools they need to create this complex chip. Most important, Meindl believes that fabricators will develop reliable new tools and techniques needed to manufacture such complex chips, mastering a new level of smallness.

So formidable are the challenges of miniaturization that chipmakers discuss the state of their art not in terms of the number of transistors, but in terms of their size. The million-element chip, or megachip, is said to use one-micrometer technology, meaning the smallest line on the chip is one micrometer wide. The gigachip must employ one-quarter-micrometer technology. The smallest lines on the face of that chip will be one four-hundredth of the diameter of a human hair—less than 500 atoms wide.

For features as tiny as these to be created on a wafer of silicon or other semiconductor, the photographic technology used to pattern the features must advance more dramatically than it has since first applied to the making of transistors in the mid-1950s. Although the details of the technique have changed many times, its essence has not. The silicon is covered with a photosensitive substance called a resist. A stencil-like mask containing the chip's pattern is placed over the surface. Then an intense light is shone through the mask to project the pattern onto the resist. A chemical washes away the resist in the pattern area, exposing it for the next fabrication step.

When the features of the projected pattern are very fine, their sharpness, or resolution, is determined in large measure by the wavelength of the light used to make the exposure. The shorter the wavelength, the better the resolution. The theoretical rule of thumb is that the wavelength must be no more than one-half the size of the smallest feature. As a practical matter, chipmakers must cut that tolerance by half again. Thus, visible light, with a wavelength of about one-half micrometer, will resolve two-micrometer lines. To pattern one-micrometer lines, fabricators use ultraviolet light having a wavelength of one-fifth micrometer. To achieve the resolution required by submi-

A human hair, enlarged about 2,000 times in the drawing below, looms gigantic in comparison to the three dots that represent the size of features on computer chips. At this scale, the left-hand dot is equivalent to one micrometer, or about one-hundredth the thickness of a hair; the narrowest features on the most densely packed chips are about one micrometer across. Dust particles of only half a micrometer *(middle dot)* can wreak havoc during chip fabrication. In the future, x-ray beams will produce chip features as small as one-tenth of a micrometer *(right-hand dot)*.

crometer resolution, they are turning to the ultrashort wavelengths of x-rays and to beams of electrons that would trace patterns directly onto the chip's surface without the need for a mask.

X-RAY LITHOGRAPHY

Experimentation with x-rays began long before photographic resolution became an issue. Because x-rays pass right through the dust and other clean-room contamination that accounts for most chip defects, fabricators looked at them as a possible route to higher yields. But not all x-rays are useful. The so-called hard x-rays used by doctors and dentists, while easy and inexpensive to produce, zip through the photoresist without exposing it. Soft x-rays—whose photons have less energy than those of hard rays—will expose the resist, but because the rays are not highly concentrated, the time necessary to record a pattern on the resist is too long for soft x-rays to be of commercial use. Only fifteen to twenty-five wafers can be exposed hourly by x-ray lithography, compared with fifty or more using an ultraviolet step-and-repeat camera. In order to use x-rays, chipmakers

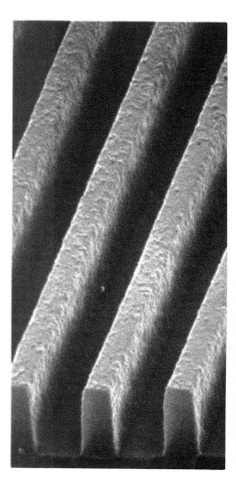

KEEPING CHIP FEATURES SHARP

As chipmakers strive to create ever-smaller chip components, they must pay close attention to the wavelength of the light used to project mask patterns onto a chip: If the wavelength is not significantly shorter than the width of the smallest mask feature, the projected image will be blurred and the resulting circuitry will malfunction. In the photograph at far left, channels developed into a layer of photoresist have a width of 0.6 micrometers. The channels are sharply defined because the wavelength of the ul- traviolet light that was used to produce them was 0.365 microme- ters. But when shone through features 0.4 micrometers wide, the same light source produced a less precise pattern that is unusable. To achieve crisp detail at even smaller sizes, researchers have turned to x-rays, using wavelengths as short as .001 micrometers. Features only .085 micrometers wide *(above)* are sharp when made with x-rays.

must either content themselves with limited production speed or invest in a better source of soft x-rays. One alternative source is a device called the synchrotron, a close cousin of the giant particle accelerators used in atomic research. The synchrotron has the unique ability to produce soft radiation that is 20 to 100 times more powerful than conventional sources.

Ordinarily, soft x-rays are created by bombarding an aluminum target with a stream of low-speed electrons from a conventional electron gun usually used for mask fabrication. The electrons, colliding with the aluminum, cause the target's electrons to emit x-rays. A synchrotron, however, uses powerful magnets to accelerate electrons almost to the speed of light and send them in a narrow beam around a racetrack pattern. Where the racetrack curves, the electrons emit radiation in the form of very intense x-rays. Fabricators can tap the energy by locating processing stations along the curve.

Size and cost have so far kept synchrotrons out of the fabrication plant. The synchrotron at Brookhaven National Laboratory, used for chipmaking experiments by IBM and other manufacturers, is fifty feet in diameter and cost $40 million. Nevertheless, IBM is installing a twenty-foot machine at its chip factory in Fishkill, New York. German and Japanese manufacturers are trying to build tabletop synchrotrons a mere six feet in diameter.

Even with this degree of miniaturization, x-ray technology is expected to remain expensive. "The projected capital cost of a synchrotron chip factory is $500 million," says Charles H. Ferguson of M.I.T.'s Center for Technology, Policy and Industrial Development. Yet despite the price, synchrotron x-ray equipment is an attractive solution to the submicron dilemma. "Time, money and commitment rather than a substantial technological breakthrough are needed," says David Elliott, an authority on chip fabrication.

ELECTRON-BEAM EXPOSURE

Another approach may be to eliminate electromagnetic radiation altogether, along with the masks, and substitute a beam of electrons to draw a pattern directly on the resist-coated wafer. Fabricators began experimenting with electron beams in the late 1960s to create the ultrafine mask patterns needed for very-large-scale integration. By the 1980s, electron-beam exposure had become the preferred way to pattern the masks used for making chips.

Electrons, released by heating a metal rod, are focused into a single beam and aimed by magnetic fields, much as the electron beam in a television tube is directed by magnets to scan the screen from side to side. The computer that controls the system contains a map of the mask design consisting of billions of points, or pixels. As the machine scans the surface of the mask material, the program turns the beam on and off to expose only the pixels that make up the pattern. Even though electron-beam systems can expose as many as 10 million pixels a second, many minutes are consumed in the exposure of just one mask.

That the system was slow had little impact on mask making; masks were costly to produce, but they could be used hundreds of times to make many thousands of chips. However, to do away with the masks and turn the electron beam directly on the chip would be far too costly. With this approach, only five to ten wafers could be processed each hour, compared to the fifty to a hundred wafers handled by a system that uses masks.

A team of ingenious Dutch researchers have raised the hope for greater speed. They use a gold-plated stainless steel grid to split an electron beam into 1,024 individually controllable beamlets, each one-tenth micrometer in diameter. The beamlets can be switched on and off independently to expose single pixels, while all 1,024 beamlets scan the wafer surface together. The device can expose 10 billion pixels per second—the equivalent of nearly a hundred wafers per hour.

SILICON AND STEEL

If all the problems of materials, design, and fabrication were to be solved in time for a gigachip to appear by the year 2000, the rate of growth in chip complexity would exceed the rate that prevailed through much of the 1970s—an annual doubling. From the megachip of 1988, the gigachip represents a 1,000-fold increase in the number of elements. Doubling the number every year—first to two million, then to four million, eight million, and so on—would assure that the billion-element mark would be reached within ten years.

James Meindl, who has observed the semiconductor industry since the days of the earliest integrated circuit, feels that gigascale integration will become a reality right on schedule, though an additional five years may pass, while demand grows and fabricators master the technology, before the device becomes common. The question for Meindl and for other seers into the future of the chip is whether the rapid growth rate can be sustained beyond the gigachip. Most projections of the computer's future have been almost laughably inaccurate. A new breakthrough in elementary physics or in some other arena of science may keep the development curve rising ever more steeply. But when Meindl puzzles over the outlook for the chip, he sees the achievement of gigascale integration as the prelude to the onset of maturity in semiconductor computer technology.

For a picture of the future, Meindl turns to the past. "Historians have often observed," he points out, "that a commercially successful technology tends to follow an S-shaped developmental curve." At the outset, a new technology materializes slowly. As it shows promise, investment increases and development proceeds at a faster—sometimes phenomenal—rate. Production rises and the technology becomes woven into the fabric of society. Eventually, however, all the technology's easy improvements have been made, there are fewer and fewer corners that it has not penetrated, and the curve of progress levels off.

Iron, for example, was the fuel of the industrial revolution, just as silicon has been the fuel of the information revolution. During the early part of the nineteenth century, the growth in the use of iron and steel was slow, as was the growth of silicon chip production in the early 1960s. Between 1860 and 1900, American steel production doubled every four years—nearly matching the rate of growth of silicon between 1965 and 1980. Since then, however, steel production has remained nearly level.

Meindl believes that the top of the semiconductor curve may be reached by the end of the next century's first quarter. Should his prediction come to pass, the semiconductor revolution will be over, and the stage will be set for the next coup. New technologies—perhaps logic circuitry founded on the transmission of light pulses rather than electric ones, or even based on molecules of organic chemicals—promise to be a continuing source of challenges to chipmakers.

A Diversity
of Chip Vehicles

Although a newly fabricated microchip comes off the assembly line laden with powerful circuitry, several finishing steps are required before it can be plugged in and put to work in an electronic device. In this final stage of manufacturing, known as packaging, the delicate sliver of silicon is mounted in a sturdy case that protects it against dust and damage. The case, or package, also carries the wiring that connects the chip within to other chips contained in the device. Because packaging procedures and equipment are completely different from those used in chip fabrication, some chipmakers farm out this part of the work to specialized packaging firms.

As in all other aspects of semiconductor manufacturing, packaging has become more challenging as chips have grown in complexity and shrunk in size. In the 1960s and 1970s, chips were relatively simple, and most could be accommodated by an equally simple plastic case with a dozen or so metal conductors protruding from two sides. To carry electrical signals from one component to another, chips were mounted on a card with metal conductors printed on its surface. But over the past decade, as improvements in chip design and fabrication increased the functions that could be crammed onto each chip and the speed of performing them, the time it took to transmit signals between chips became a delaying factor. Package designers sought to solve this problem using the principle of compression—packing as many functions as possible onto a single chip, and then squeezing several of these densely wired chips onto the smallest circuit board possible.

The more complex the chip's circuitry, however, the more metal conductors are needed on the board, and the less room there is available for components. Here lies the packager's chief quandary: to create a package large enough to provide connections for a modern chip's dense wiring and yet small enough to fit with other packages on the board. To handle the space problem, packagers have devised such solutions as placing conductors on all four sides of a package, creating a sandwich-style circuit board with multiple layers of wiring, and even creating packages that combine functions formerly divided among many separate components on a board.

Making Connections

Bonding is the first step in packaging a chip. The purpose is to establish electrical connections between the chip and the package conductors, called leads or pins, that link the chip to a circuit board.

Wire bonding, the oldest and most widely used of the three methods shown on these pages, is preceded by attaching the chip to a framework of leads. Most chips intended for wire bonding are glued to the leads. Others, however, are manufactured with a thin backing of gold. Portions of the package leads are gold- or silver-plated. When pressed together and heated, the gold on the back of the chip fuses with the gold or silver on the leads to secure the chip.

To wire the chip to the leads, both are heated to about 200 degrees Celsius. Then a remarkable machine called a ball bonder goes to work. Guided by computer and working at a rate of five connections or more per second, it first melts the end of a fine gold wire—sometimes no thicker than fifteen micrometers in diameter—so that it forms a tiny ball. Then, as shown in the top row of illustrations at right, the machine fastens the wire to one of several square aluminum bonding pads to which the chip's internal wiring is connected. Heat, pressure—about fifty grams per square inch—and ultrasonic vibrations at frequencies up to 60,000 hertz (cycles per second) or so mix atoms of gold and aluminum where the wire touches the bonding pad. In chips that operate at high temperature, a gold-aluminum bond is unreliable, so aluminum wire replaces the gold. In bonding aluminum to aluminum, heat is unnecessary; only ultrasonic vibrations and pressure are used to assure a strong connection.

In flip-chip bonding (right, center), the process that cements the chip to the base also makes the electrical connections. Though more expensive than wire bonding, each flip-chip bonding is valued because, in eliminating the need for a long wire from lead to bonding pad, it speeds chip operation. In tape-automated bonding (right, bottom) a lacy web of conductors etched through a film of gold is first bonded to the chip. Then the chip and conductors are attached to leads. Etching the conductors allows them to be closely spaced, thereby fitting more wires into the small areas of most chips.

Wire Bonding

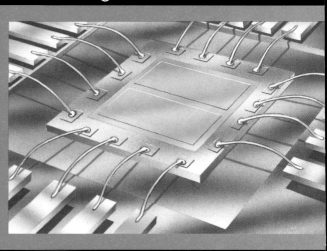

Flip-Chip Bonding

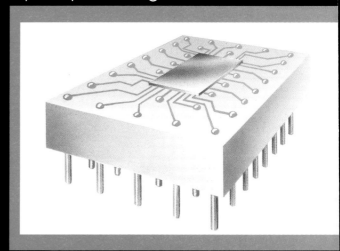

Tape-Automated Bonding

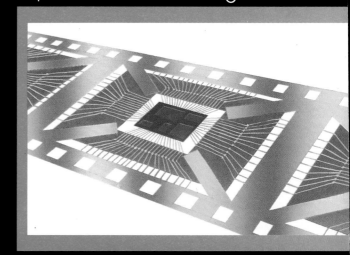

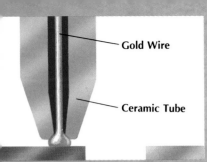

Gold Wire

Ceramic Tube

> arch wires between a chip and its
ckage *(left)*, a ceramic tube, vibrat-
g imperceptibly, presses the ball-
aped end of a gold wire onto a
eated bonding pad at the chip's edge.

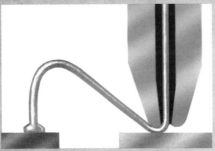

Wire from a spool passes through the
tube as it moves to the lead frame,
also heated. The wire is pressed into
place by the tube and nicked by a
sharp ridge at its tip.

After completing both bonds, the tube
ascends, breaking the wire at the nick.
The nub of wire extending from the
tube will be melted to form a gold ball
for the next connection.

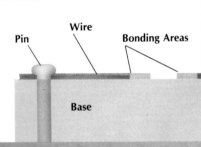

Pin
Wire
Bonding Areas

Base

inlike leads extend through the ceramic base
sed in this type of bonding. The lower ends
f the pins will plug into a circuit board; the
ops are linked by fine wires to bonding areas
n the surface of the ceramic.

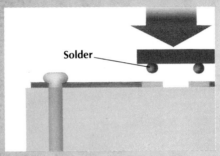

Solder

Balls of molten solder are deposited on raised
contact pads on the underside of the chip
and allowed to cool. Then the chip is set onto
the base, with the solder balls coinciding
with bonding areas.

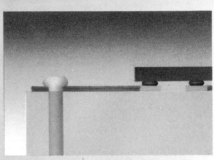

The chip and base are then heated to about
275 degrees Celsius, a temperature high
enough to melt the solder and cement the
chip to the bonding areas, but low enough
not to damage the chip.

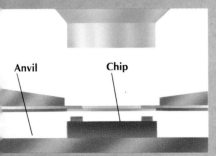

Anvil
Chip

o join a chip to gold conductors *(left)*,
t is placed on a heated platform called
n anvil *(above)*. The conductors are
ositioned above the chip by means of
procket holes in a tape support.

Press

As a ceramic press descends toward
the tape, the anvil rises, pressing
the conductors and the chip together.
Heat and pressure stick the chip
to the conductors.

After the bond is made, the press and
anvil retract. The tape lifts the bonded
chip away from the anvil and then
advances, moving the next set of conductors into position.

A Package
for Every Purpose

Designs and materials used for chip packaging must meet several practical requirements. The casing material should dissipate heat, create a tight seal, and not interfere with the chip's electrical activity. Leads must be good electrical conductors and rugged enough to withstand the rigors of being attached by automated machinery to a circuit board. And the completed package should be no larger than absolutely necessary, in order to fit the most chips into the smallest circuit-board space possible.

Plastic is the most common casing material, because it is the least expensive. More than 80 percent of the world's integrated circuits are housed in what are known as dual in-line packages (DIPs) using a method called plastic injection molding *(right, top)*. To create this kind of casing, the chip and its leads are placed in a mold, which is then filled with hot plastic that engulfs the chip and forms a tight seal upon cooling. While this type of casing is satisfactory for many components, a plastic package is vulnerable to corrosion-causing water

vapor, which can enter through gaps that are sometimes left around leads, and plastic is slow at dissipating the heat generated by the dense circuitry of many chips. Instead of plastic, packagers sometimes use a ceramic casing material. Though costlier, the ceramic provides a moistureproof seal and also siphons off heat more effectively than plastic does.

As shown below, packages come in a wide variety of lead arrangements and are mounted on circuit boards in one of three ways. With pin-in-hole mounting, the oldest method of attaching packages, metal leads protruding from the casing are inserted through holes in the circuit board to connect with conductors on the underside. A space-saving alternative called surface mounting eliminates the holes; leads are soldered directly to conductors on one side of the board, leaving the other side free for additional chips and conductors. A third method involves inserting the package's lead into a socket attached to the board's surface, a convenience for devices that might later be removed from the board.

Provision for Connections

A PACKAGE
FOR SIMPLE CHIPS

This tiny DIP was developed by the Swiss for electronic watches. Because of the package's small number of leads, it is primarily used for simple logic or memory chips, which require few electrical connections. This package is usually surface mounted using gull-wing leads, which are named for their distinctive curve.

A PACKAGE DESIGNED
TO SAVE SPACE

This pin-in-hole package is mounted on edge to save circuit-board space. With a single row of leads, it is often used for memory chips—which individually require few connections—so that the packages can be packed together to form a powerful memory bank.

A ZIGZAG PATTERN
OF LEADS

An additional lead or two can fit along the edge of a package if the leads are laid out in two rows. Leads in one row are positioned opposite gaps between leads in the other row to keep them separated from one another. Bends in the leads permit holes through the circuit board to be drilled far enough apart not to overlap.

A Chip in Its Package

A cutaway view of a common type of package, called a dual in-line package (DIP) because of the two rows of leads extending from the sides, reveals the tiny chip in the center. Fine wires connect the chip to a pattern of metal leads, which provide electrical conduits to a circuit board.

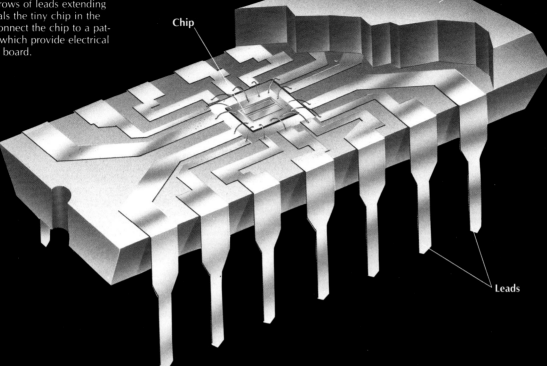

Plastic Case

Chip

Leads

3

A LEADLESS CERAMIC PACKAGE

Placing leads along all four sides of a casing increases the number of electrical connections without using additional board space. In this version, conductors make contact with leads at the sides of a ceramic base then wrap around to the bottom for surface mounting or inserting into a socket.

GULL WINGS FOR RELIABILITY

A plastic package is often unsuited to a leadless design; as a chip warms and cools even slightly, the plastic would expand and contract enough to crack solder joints to the circuit board. To solve the problem, packagers use gull-wing leads that flex upon heating or cooling to relieve stress on soldered connections.

J-LEADS FOR COMPACTNESS

In a variation of the gull-wing design, leads of this surface-mounted package curve inward. J-leads protect solder joints as effectively as gull-wing leads. By curling under the package, they also reduce circuit-board space occupied by a package and allow chips to be mounted closer together.

A PIN-GRID ARRAY FOR MORE LEADS

This design, which has a ceramic base penetrated by a grid of fine, closely spaced leads, permits nearly the entire underside of the package to be used for electrical connections. Because leads pass through a circuit board, this package is reserved for complex microprocessors and other complex chips that need more leads than would fit around the edges of surface-mount packages of equal area.

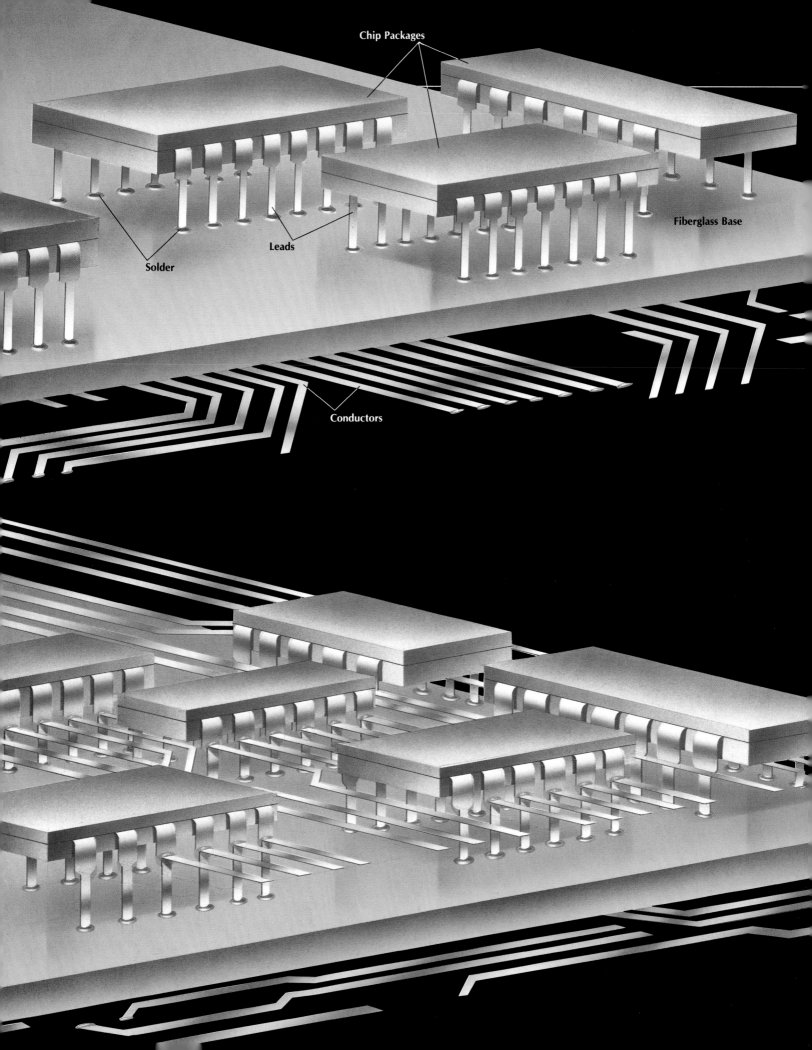

Chip Packages

Fiberglass Base

Solder

Leads

Conductors

Layers of Conductors

A Single-Layer Board

In this diagram of a single-layer circuit board, pin-in-hole chip packages are plugged into the board's fiberglass base *(green)*. Package leads extend through holes drilled in the base and connect with copper conductors on the underside. For clarity, the conductors are shown here pulled away from the base. A dot of solder at the end of each conductor indicates where a lead makes electrical contact with the copper.

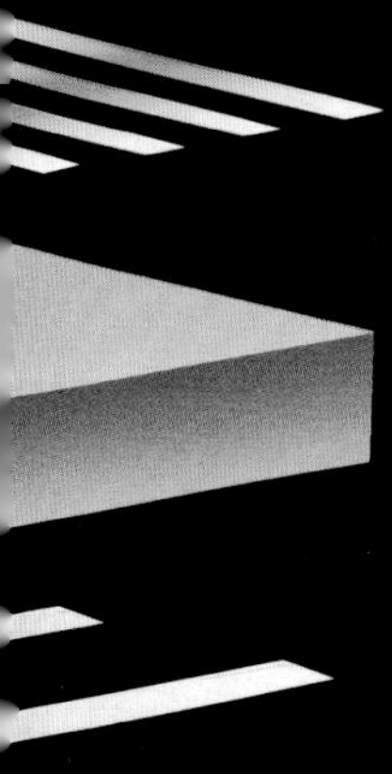

A printed circuit board supports electronic components mounted on its surface and links them electrically. It consists of a nonconductive base, typically of fiberglass, coated with a thin layer of metal such as copper, which is etched to form a pattern of electrical conductors.

The simplest boards have only one layer of wiring, often applied on the bottom *(left, top)*. Adding conductors to the top surface *(left, bottom)* makes it possible to connect many more components on a board. Many computer circuits have become so complicated that they require as many as a dozen layers of conductors, sandwiched between thin insulating layers of fiberglass, in order to make the necessary connections *(overleaf)*.

Multiple layers of conductors also make it possible for electrical paths to cross without causing a short circuit. Wires in each layer run roughly parallel to each other, and at a right angle to the wiring in layers above and below. A signal that must change direction en route to its destination can move up or down through insulating fiberglass to pick up the next copper leg of its journey.

The first step in preparing a board is to create a photographic transparency of the circuitry layout, in which the pattern of conductors is black and opaque to light. Next, the transparency is projected onto the board after the copper has been coated with a film of photoresist that hardens on exposure to light. Washing away the soft parts of the film in a chemical bath reveals the pattern of conductors. This tracery is coated with solder to protect the conductors when the rest of the photoresist and the copper that it shielded from the plating process are washed away.

A Double-Layer Board

The circuit board shown at left has copper conductors on both sides. Conductors in the upper layer travel lengthwise along the board, while those in the lower layer run across it. Pin-in-hole packages are mounted on one side; some leads connect through to wires on the bottom layer, while others make contact on the top. Though none are illustrated, surface-mounted components can be soldered to either side of the board.

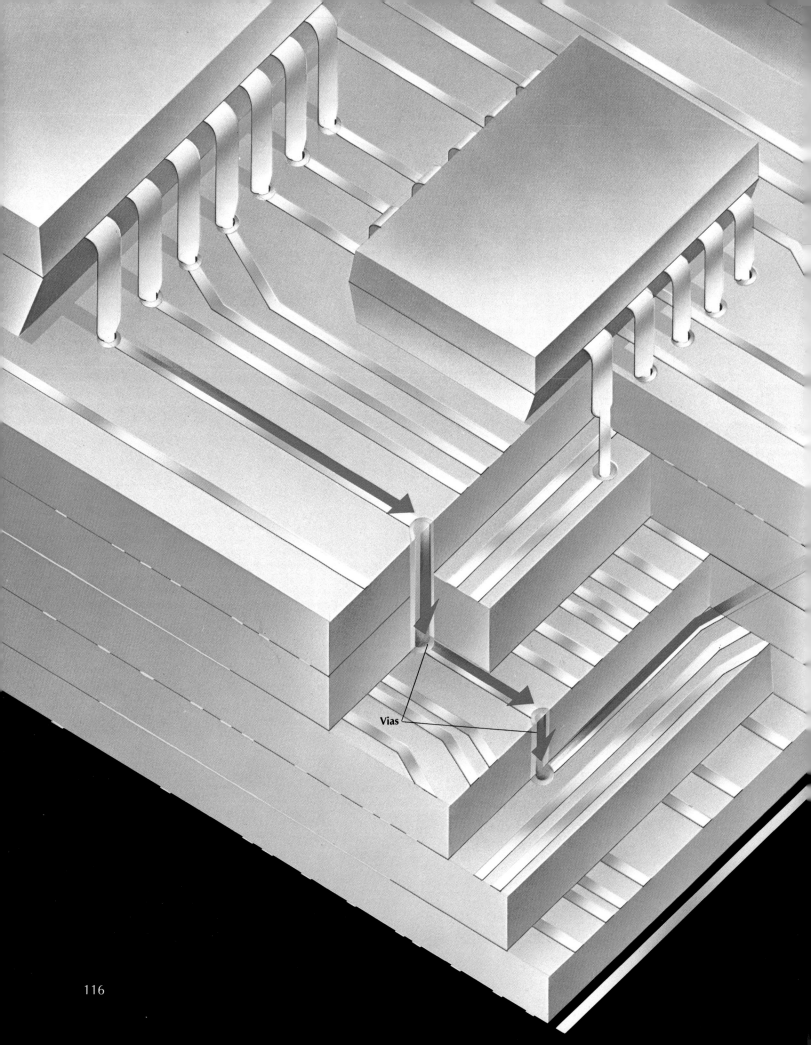

Vias

Via

Conductors

Conductive shafts, called vias, are drilled through the board after it is assembled. For a hole to carry electricity, the sides must be coated with copper by a process that also deposits metal on the surface of a board. To prevent copper from filling the space between conductors of the outermost layers, the shafts are drilled and coppered before these conductors are etched. A via at its simplest connects conductors on just two levels, the top and the fourth in this instance. Conductors on other levels stop short of the via and jog around it.

A Multilayered Maze

Six levels of circuitry in this circuit board permit a pulse of data *(red arrow)* to reach its destination by burrowing under obstacles. The route first leads the pulse across the board on the top layer. Unable to link directly with the destination without crossing other conductors, the route then descends by means of a conductive hole called a via *(above, left)* to a layer of pathways that travel lengthwise. The first such level is unusable; circuitry for another chip already occupies the space. The path must descend through the third layer before reaching wiring on the fourth level that provides a clear route to its destination.

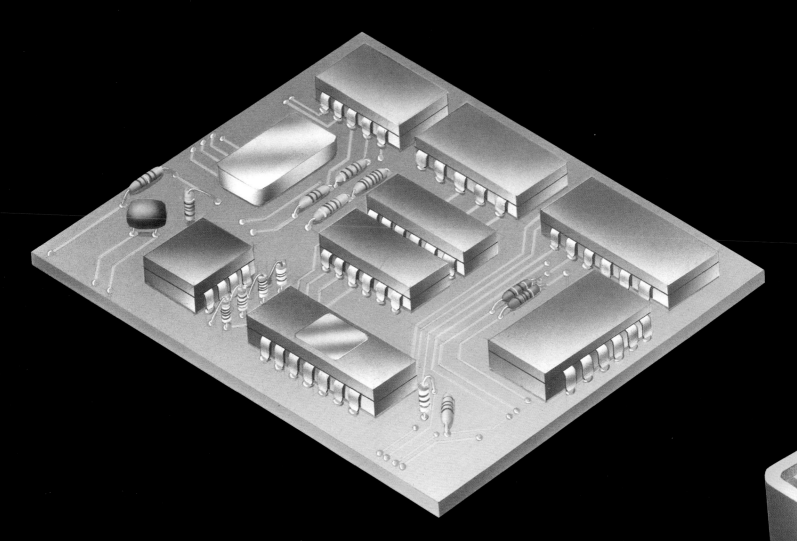

One Chip per Package

This small circuit board, used in the early 1970s to monitor the rotary motion of an airplane's flaps, ailerons, and rudder, was roughly two and three-quarters inches long by two inches wide. The board had numerous simple microchips mounted on it along with many other components necessary to report control-surface movement to the pilot.

The Disappearing Circuit Board

The circuit board, with its array of individual chips and elaborate interconnections, is hardly a model of efficiency. It is costly to build, and the considerable distances that data must travel between chips slow computer operations.

To speed things up—and lower the price—entire circuit boards of chips have been removed from their separate packages and rearranged on a board small enough to fit into a single package called a hybrid. If a hybrid is in great demand, it becomes economical to shrink the circuitry even further by re-creating it in a single piece of silicon, a development that parallels the steady miniaturization of microcircuitry.

The result has been to vastly reduce the dimensions and even the number of circuit boards necessary in electronic devices, from the microprocessor nerve center of a computer to the circuitry shown on these pages, which reports to an airplane pilot the positions of his aircraft's control surfaces.

A Miniature Circuit Board

By the late 1970s, the functions of the circuit board had been compressed onto fewer and smaller components and repackaged as a hybrid, about one and one-half inches by three-quarters of an inch in size. The hybrid, shown here without its protective lid, consists of various chips mounted on a ceramic base and connected just as the earlier devices were arranged on a circuit board.

The Works on One Chip

A decade later, the circuitry needed to perform the same functions as the circuit board and hybrid had been squeezed onto a single quarter-inch-square chip. Though the package, shown with its protective cover removed, is little smaller than the hybrid, the circuitry is compressed into a significantly smaller space, making for much faster computing at about one-fifth the cost of the original board.

Glossary

Algorithm: a set of clearly defined rules and instructions for the solution of a problem. Before a computer can solve a problem, it must be given a carefully planned algorithm.

Alignment: in chip manufacturing, the arranging of the mask and the wafer, one with respect to the other, usually followed by exposure to light.

Analog: continuous representation of a changing variable by another physical variable; for example, temperature as the height of a column of mercury.

Base: region of a bipolar transistor that controls the flow of current between the emitter and the collector.

Bias: a voltage applied to a device to establish a reference level for operation.

Bipolar: transistors in which current is carried by both positive and negative particles, holes, and electrons.

Bit: a binary digit. A bit is the smallest unit of storage in a digital computer and is represented by a zero or a one.

Bus: a circuit or group of circuits that provides communication between electronic components.

Capacitance: the amount of charge that a capacitor can store.

Capacitor: a device that stores electrical energy when a voltage is applied. It consists of two conductors separated by an insulator called a dielectric.

Carriers: holes or electrons that are available in a semiconductor substrate for conduction of electric current.

Channel: the region between the source and drain of a field-effect transistor through which current flows.

Chemical vapor deposition (CVD): a method of depositing insulators, semiconductors, and metals in which a chemical reaction of gases causes the desired material to precipitate onto the surface of the wafer.

Chip: integrated circuit.

Chip carrier: a variety of package.

Circuit board: the board, often fiberglass, on which electronic components are mounted.

Collector: section of a bipolar transistor that attracts current emanating from the emitter.

Complementary metal-oxide semiconductor (CMOS): a logic circuit created by combining negative-channel and positive-channel metal-oxide semiconductor transistors.

Contact: during chip fabrication, the area of silicon that is covered with metal to serve as a point of electrical access to the chip.

Contamination: foreign material—a dust particle, for example—that adversely affects the physical or electrical characteristics of a wafer.

Die: *See* Chip.

Dielectric: an insulator.

Diode: an electronic device that allows current to flow in only one direction.

Discrete component: a separate component, such as a resistor or capacitor, used in a conventional circuit. Components on a chip are integrated, rather than discrete.

Doping: deliberately introducing impurities into a semiconductor to alter its conductivity.

Drain: in MOS transistors, the area toward which current flows when the switch is turned on; equivalent to the collector in a bipolar transistor.

Dual in-line package (DIP): the most common type of chip package. Electrical connectors called leads or pins extend downward from opposite sides of this package's rectangular body.

Dynamic random-access memory chip (DRAM): an integrated circuit that serves as temporary memory in computers.

Electron: a negatively charged particle revolving around the nucleus of an atom.

Emitter: in a bipolar transistor, the region of semiconductor material that supplies current-carrying electrons or holes.

Emitter-coupled logic (ECL): a bipolar technology generally used where very high switching speeds are required.

Epitaxial: having the same structure as an underlying layer. Used to describe the process of growing a crystal layer upon another crystal so that the new crystal has the same crystalline structure as the one on which it is grown.

Etching: chemical removal of material (such as silicon dioxide) in a defined area of a chip.

Field-effect transistor (FET): a solid-state device in which current is controlled between source terminal and drain terminal by voltage applied to a nonconducting gate terminal.

Flip-flop: an electrical circuit with two stable states—on or off; the basic component of a logic or memory circuit.

Gallium arsenide: a synthetic semiconductor used in the construction of high-speed transistors and integrated circuits; valued for its low power consumption and ability to withstand heat.

Gate: (a) a circuit that accepts one or more inputs and always produces a single predictable output; (b) the control terminal in a MOS transistor.

Gate array: an integrated circuit consisting of a regular arrangement of gates (a) that can be interconnected to perform specific functions.

Hole: the concept used to describe the movement of the absence of an electron through the crystal structure of a semiconductor; the movement of a hole is equivalent to the movement of a positive charge.

Hybrid: an assembly of one or more chips mounted on a base, often ceramic, connected to each other by wiring printed on the base, and enclosed in a single chip package.

Ingot: in chip fabrication, a single large silicon crystal several inches in diameter and up to six feet long grown around a tiny seed crystal. The ingot is then sliced into wafers, polished, and used to fabricate chips.

Integrated circuit: an electronic circuit whose components are all formed on a single piece of semiconductor material, usually silicon; a chip.

Ion: an atom that has either gained or lost electrons, making it either a positively or negatively charged particle.

Ion implantation: process used to drive selected impurities into a semiconductor in order to achieve desired electrical properties in selected regions.

Junction: on a semiconductor, the boundary between a p-type region and an n-type one.

Large-scale integration (LSI): the placement of 100 to 5,000 gate equivalents, or 1,000 to 16,000 bits of memory, on a single chip.

Lithography: the transfer of a design from one medium to another, for example, from a mask to a wafer. If light is used to perform the transfer, the process is called photolithography.

Mask: a patterned plate, such as a stencil, through which light is projected to define areas of a chip to be processed.

Metallization: deposition of a uniform film of metal over the surface of a wafer, interconnecting the various elements of each integrated circuit on that wafer.

Metal-oxide field-effect transistor (MOSFET): a variety of field-effect transistor.

Metal-oxide semiconductor (MOS): a technology for constructing integrated circuits with layers of conducting metal, semiconductor material, and silicon dioxide as an insulator.

Micrometer: one millionth of a meter.

Mobility: a measure of the speed with which charges move through a semiconductor.

Monolithic device: one having circuitry that is completely contained on a single chip.

N-type semiconductor: one through which current flows primarily by electrons.

Oxide masking: use of an oxide on a semiconductor to create a pattern for doping.

Package: case used for physically protecting and securing a chip and electrically connecting it to other electronic components.

Passivation: treatment of a region of a chip to protect electronic properties from chemical action or corrosion.

Photoresist: light-sensitive material deposited as a thin film on silicon wafers and used as a stencil for etching or doping.

Polycrystalline silicon: silicon made up of many jumbled crystals. Raw silicon comes in polycrystalline form prior to melting and growth of the single crystal necessary for semiconductor fabrication.

P-type semiconductor: one through which current flows primarily by holes.

Semiconductor: a material that has properties of both a conductor and an insulator. Common semiconductors include germanium and silicon.

Small-scale integration (SSI): fewer than ten logic gates on a chip.

Solid state: pertaining to electronic devices, such as transistors, made from a semiconductor.

Source: in a MOS transistor, the area of semiconductor material that supplies current-carrying electrons; equivalent to the emitter in a bipolar transistor.

Standard cells: predefined logic elements consisting of a single transistor or groups of transistors used to speed design of an integrated circuit.

Transistor: the basic solid-state device used to amplify or switch current.

Unipolar transistor: one through which current is carried either by electrons or by holes, but not by both.

Very-large-scale integration (VLSI): the placement of 5,000 or more gate equivalents or more than 16,000 bits of memory on a single chip.

Wafer: a thin disk of semiconductor material, usually silicon, on which many separate chips can be fabricated and then cut into individual units.

Yield: the number of units produced compared to the maximum number possible. In chip fabrication, a wafer's yield is referred to as the number of chips that actually work compared to the number the wafer holds.

Bibliography

Books

Augarten, Stan:
  Bit by Bit. New York: Ticknor & Fields, 1984.
  State of the Art: A Photographic History of the Integrated Circuit. New Haven, Conn.: Ticknor & Fields, 1983.
Bernstein, Jeremy, *Three Degrees above Zero: Bell Labs in the Information Age.* New York: Charles Scribner's Sons, 1984.
Bilotta, Anthony J., *Connections in Electronic Assemblies.* New York: Marcel Dekker, Inc., 1985.
Braun, Ernest, and Stuart MacDonald, *Revolution in Miniature.* Cambridge: Cambridge University Press, 1982.
Bylinsky, Gene, *High Tech: Window to the Future.* Hong Kong: Intercontinental Publishing Corporation Ltd., 1985.
Caddes, Carolyn, *Portraits of Success: Impressions of Silicon Valley Pioneers.* Palo Alto, Calif.: Tioga Publishing Company, 1986.
Computer Basics, by the Editors of Time-Life Books (Understanding Computers series). Alexandria, Va.: Time-Life Books, 1985.
Davies, Helen, and Mike Wharton, *Inside the Chip.* London: Usborne Publishing Ltd., 1983.
Elliott, David J.:
  Integrated Circuit Fabrication Technology. New York: McGraw-Hill Book Company, 1982.
  Integrated Circuit Mask Technology. New York: McGraw-Hill Book Company, 1985.
  Microlithography: Process Technology for IC Fabrication. New York: McGraw-Hill Book Company, 1986.
Handel, Samuel, *The Electronic Revolution.* Baltimore, Md.: Penguin Books, 1967.
Hanson, Dirk, *The New Alchemists.* Boston: Little, Brown and Company, 1982.
Harper, Charles A., ed., *Handbook of Electronic Packaging.* New York: McGraw-Hill Book Company, 1969.
Hnatek, Eugene R., *A User's Handbook of Integrated Circuits.* New York: John Wiley & Sons, 1973.
Jackson, Peter, *The Chip.* New York: Warwick Press, 1986.
Jaeger, Richard C., *Introduction to Microelectronic Fabrication.* Vol. 5 of *Modular Series on Solid State Devices.* Reading, Mass.: Addison-Wesley Publishing Company, 1988.

Malone, Michael S., *The Big Score: The Billion-Dollar Story of Silicon Valley.* Garden City, N.Y.: Doubleday & Company, Inc., 1985.

Maly, W., *Atlas of IC Technologies: An Introduction to VLSI Processes.* Menlo Park, Calif.: The Benjamin/Cummings Publishing Company, Inc., 1987.

Mead, Carver, and Lynn Conway, *Introduction to VLSI Systems.* Reading, Mass.: Addison-Wesley Publishing Company, 1980.

Millman, Jacob, *Microelectronics: Digital and Analog Circuits and Systems.* New York: McGraw-Hill Book Company, 1979.

Millman, S., ed., *A History of Engineering and Science in the Bell System: Physical Sciences (1925-1980).* Murray Hill, N.J.: AT&T Bell Laboratories, 1983.

Pletsch, Bill, *Integrated Circuits: Making the Miracle Chip.* Portland, Oreg.: Irwin Hodson Co., 1978.

Powell, R. A., *Welcome to Lilliput: A Microelectronics Overview.* Palo Alto, Calif.: Varian Associates, Inc., 1985.

Queisser, Hans, *The Conquest of the Microchip.* Cambridge: Harvard University Press, 1988.

Ralston, Anthony, and Edwin D. Reilly, Jr., eds., *Encyclopedia of Computer Science and Engineering.* New York: Van Nostrand Reinhold Company, 1983.

Reid, T. R., *The Chip.* New York: Simon & Schuster, 1984.

Rice, Rex, ed., *VLSI Support Technologies: Computer-Aided Design, Testing and Packaging.* Los Alamitos, Calif.: IEEE Computer Society Press, 1982.

Rogers, Everett M., and Judith K. Larsen, *Silicon Valley Fever: Growth of High-Technology Culture.* New York: Basic Books, Inc., 1984.

Sanders, Donald H., *Computers Today.* New York: McGraw-Hill Book Company, 1983.

Shurkin, Joel, *Engines of the Mind: A History of the Computer.* New York: Simon & Schuster, Inc., 1984.

Simon, John J., Jr., ed., *From Sand to Circuits and Other Inquiries.* Cambridge: Harvard University Office for Information Technology, 1986.

Van Zant, Peter, *Microchip Fabrication.* San Jose, Calif.: Semiconductor Services, 1985.

Walton, Mary, *The Deming Management Method.* New York: Dodd, Mead & Company, 1986.

Warner, R. M., Jr., and B. L. Grung, *Transistors: Fundamentals for the Integrated-Circuit Engineer.* New York: John Wiley & Sons, Inc., 1983.

Weste, Neil H. E., and Kamran Eshraghian, *Principles of CMOS VLSI Design: A Systems Perspective.* Reading, Mass.: Addison-Wesley Publishing Company, 1985.

Wolf, Stanley, and Richard N. Tauber, *Silicon Processing for the VLSI Era, Vol. 1: Process Technology.* Sunset Beach, Calif.: Lattice Press, 1986.

Zaks, Rodnay, and Alexander Wolfe, *From Chips to Systems: An Introduction to Microcomputers.* Alameda, Calif.: SYBEX Inc., 1987.

Periodicals

Abelson, Philip H., and Allen L. Hammond, "The Electronics Revolution." *Science,* March 18, 1977.

Abraham, Howard E., et al., "NMOS-III Photolithography." *Hewlett-Packard Journal,* August 1983.

Alper, Joseph, "Industry's New Magic Lantern." *High Technology,* April 1984.

Bairstow, Jeffrey:
   "Can the U.S. Semiconductor Industry Be Saved?" *High Technology,* May 1987.
   "Silicon Done Your Way." *High Technology,* November 1986.

Bell, C. Gordon, "A Surge for Solid State." *IEEE Spectrum,* April 1986.

Belter, Stephen E., "Computer-Aided Routing of Printed Circuit Boards." *BYTE,* June 1987.

Blodgett, Albert J., Jr., "Microelectronic Packaging." *Scientific American,* July 1983.

Boraiko, Allen A., "The Chip." *National Geographic,* October 1982.

Brandt, Richard, "The Peaks and Valleys Are Leveling Out." *Business Week,* January 11, 1988.

Broad, William J., "Quest for Fastest Computer Chip: International Contest Intensifies." *The New York Times,* February 23, 1988.

Brody, Herb, "The Bumpy Road to Submicron Lithography." *High Technology,* March 1983.

Broers, A. N., and M. Hatzakis, "Microcircuits by Electron Beam." *Scientific American,* November 1972.

Buehler, Martin G., and Loren W. Linholm, "Role of Test Chips in Coordinating Logic and Circuit Design and Layout Aids for VLSI." *Solid State Technology,* September 1981.

Burgess, Robert M., Kathleen B. Koens, and Emil M. Pignetti, Jr., "Semiconductor Final Test Logistics and Product Dispositioning Systems." *IBM Journal of Research and Development,* September 1982.

Burggraaf, Pieter, "Forecasting for a Competitive Edge." *Semiconductor International,* January 1988.

Byers, T. J., "A Revolution in IC Packaging." *Radio Electronics,* May 1986.

Bylinsky, Gene, "What's Sexier and Speedier Than Silicon." *Fortune,* June 24, 1985.

Carver, G. P., L. W. Linholm, and T. J. Russell, "Use of Microelectronic Test Structures to Characterize IC Materials, Processes, and Processing Equipment." *Solid State Technology,* September 1980.

Cook, Rick, "Chip Makers Try For Fewer Duds." *High Technology,* January 1987.

Corcoran, Elizabeth, "She Incites Revolutions with Chips and Networks." *IEEE Spectrum,* December 1987.

"The Datamation Hall of Fame." *Datamation,* September 16, 1987.

Davis, Stephen G., "The Superconductive Computer in Your Future." *Datamation,* August 15, 1987.

Factor, H., and K. Kaufman, "Wafer Dicing: On the Threshold of Automation." *Solid State Technology,* July 1985.

"Fastest Transistors in the World." *Science News,* October 18, 1986.

Feibus, Michael, and Alex Barnum, "Chipmakers Clean House: They Have Met the Enemy and It Is Dust." *San Jose Mercury*

News, December 7, 1987.

Ferguson, Charles H., "From the People Who Brought You Voodoo Economics." *Harvard Business Review,* May-June 1988.

Fierman, Jaclyn, "Automatic Eyes to Spy Chipmakers' Errors." *Fortune,* December 24, 1984.

Foner, Simon, and Terry P. Orlando, "Superconductors: The Long Road Ahead." *Technology Review,* March 1988.

Fraust, C. L., "Semiconductor Workers Face Unique Health Risk in Manufacturing Sector." *Occupational Health & Safety,* October 1987.

"Gallium Arsenide, The Supersemiconductor." *Cray Channels,* winter 1986.

"Getting the Bugs Out of Packaging." *Science News,* March 21, 1987.

Gilder, George:
"The Revitalization of Everything: The Law of the Microcosm." *Harvard Business Review,* April 1988.
"You Ain't Seen Nothing Yet." *Forbes,* April 4, 1988.

Gleick, James, "In Trenches of Science." *New York Times Magazine,* August 16, 1987.

Grossblatt, Robert, "The Evolution of VHSIC." *Radio Electronics,* March 1987.

Harris, Karl L., Paul Sandland, and Russell M. Singleton:
"Automated Inspection of Wafer Patterns with Applications in Stepping, Projection and Direct-Write Lithography." *Solid State Technology,* February 1984.
"Wafer Inspection Automation: Current and Future Needs." *Solid State Technology,* August 1983.

Helgeland, Walter, Carl Chartier, Jon Talbott, and Fred Meier, "Czochralski Crystal Growth into the '90s." *Microelectronic Manufacturing and Testing,* May 1986.

Helgeland, Walter, Ken Kerwin, and Carl Chartier, "Growing Semiconductor Crystals Using the Czochralski Method." *Microelectronic Manufacturing and Testing,* April 1983.

Heppenheimer, T. A., "Nerves of Silicon." *Discover,* February 1988.

Hittinger, William C., "Metal-Oxide-Semiconductor Technology." *Scientific American,* August 1973.

Hittinger, William C., and Morgan Sparks, "Microelectronics." *Scientific American,* November 1965.

Holton, William C., "The Large-Scale Integration of Microelectronic Circuits." *Scientific American,* September 1977.

Hong, Se June, and Ravi Nair, "Wire-Routing Machines—New Tools for VLSI Physical Design." *Proceedings of the IEEE,* January 1983.

Johnsen, Gregory, "Gallium Arsenide Chips Emerge from the Lab." *High Technology,* July 1984.

Kahng, Dawon, "A Historical Perspective on the Development of MOS Transistors and Related Devices." *IEEE Transactions on Electron Devices,* July 1976.

Kilby, Jack S., "Invention of the Integrated Circuit." *IEEE Transactions on Electron Devices,* July 1976.

Lim, Biron, "Better Chips through Chemical Vapors." *High Technology,* February 1986.

Linvill, John G., and C. Lester Hogan, "Intellectual and Economic Fuel for the Electronics Revolution." *Science,* March 18, 1977.

McBride, James, "Keyboard Contenders. Ensoniq: A U.S. Firm Takes on the Japanese." *The Washington Post,* January 6, 1988.

Main, Jeremy:
"The Curmudgeon Who Talks Tough on Quality." *Fortune,* June 25, 1984.
"Under the Spell of the Quality Gurus." *Fortune,* August 18, 1986.

Maranto, Gina, "Superconductivity: Hype vs. Reality." *Discover,* August 1987.

Marshall, Martin, Larry Waller, and Howard Wolff, "The 1981 Award for Achievement." *Electronics,* October 20, 1981.

Mazor, Itzik, and Haim Factor, "In-Line Quality Control: Key to Fully Automated Wafer Dicing." *Microelectronic Manufacturing and Testing,* March 1986.

Meindl, James D.:
"Chips for Advanced Computing." *Scientific American,* October 1987.
"Microelectronic Circuit Elements." *Scientific American,* September 1977.

Nicholls, Bill, "Systems Implications of the Intel 80386." *BYTE,* 1986 special issue.

Niewierski, Walter J., "8- and 16-bit Processors Round Out High-Level CMOS Architecture Options." *Electronics,* April 5, 1984.

Noyce, Robert N., "Microelectronics." *Scientific American,* September 1977.

Oldham, William G., "The Fabrication of Microelectronic Circuits." *Scientific American,* September 1977.

"Our Life has Changed." *Business Week,* April 6, 1987.

Perry, Tekla S., and Paul Wallich, "Inside the PARC: The 'Information Architects.'" *IEEE Spectrum,* October 1985.

Port, Otis:
"How to Make It Right the First Time." *Business Week,* June 8, 1987.
"Special Report: The Push for Quality." *Business Week,* June 8, 1987.

Posa, John G., "Superchips Face Design Challenge." *High Technology,* January 1983.

Potts, Mark, "U.S. Chip Industry's Gloomy Future." *The Washington Post,* May 10, 1987.

Reid, T. R., "The Texas Edison." *Texas Monthly,* July 1982.

Robinson, Arthur L., "One Billion Transistors on a Chip?" *Science,* January 20, 1984.

Rothenberg, Marian S., "Bell Labs Spinoffs." *High Technology,* June 1987.

Sanger, David E., "Chip Designers Seek Haste without Waste." *The New York Times,* January 6, 1988.

Schrage, Michael, "Silicon Sees New Uses: Chip Mimics Eye." *The Washington Post,* April 20, 1986.

Schroen, Walter, "Chip Packages Enter the 21st Century." *Machine Design,* February 11, 1988.

Singer, Peter H., "The Transistor: 40 Years Later." *Semiconductor International,* January 1988.

"Speeding to a Gallium Arsenide Record." *Science News,* January 10, 1986.

"Superconductivity Clears Power-Capacity Hurdle." *Wall Street Journal,* May 24, 1988.

"Superconductors." *Time,* May 11, 1987.

Taguchi, Genichi, "How Japan Defines Quality." *Design News,* July 8, 1985.

"T.I. Contributes Historic Exhibit to Science Museum Exhibition." *TIdings,* May 1980.

"TI Process Boosts Linear CMOS ICs to LSI Densities." *Electronics,* December 18, 1986.

"The Transistor." *Bell Laboratories Record,* August 1948.

"The Transistor: The First 40 Years." *Solid State Technology,* December 1987.

"Turning a PC into a Silicon Compiler." *Electronics,* June 16, 1986.

Wallich, Paul, ed., "U.S. Semiconductor Industry: Getting It Together." *IEEE Spectrum,* April 1986.

Wells, Paul, "Intel's 80386 Architecture." *BYTE,* 1986 special issue.

"Why IBM Gave Up on the Josephson Junction." *Business Week,* November 21, 1983.

Woisard, Margaret, "Cleaner Chemicals for Chip Processing." *High Technology,* April 1987.

Wollard, Katherine, "Solid State." *IEEE Spectrum,* January 1988.

Other Publications

"Building for the Future." Pamphlet. Redondo Beach, Calif.: TRW Electronic Systems Group, 1985.

Covert, Colin, "Chip Shots." *TWA Ambassador,* November 1983.

Deming, W. Edwards, and Raymond T. Birge, "On the Statistical Theory of Errors," *Reviews of Modern Physics,* July 1934.

"8300 Design Criteria for Low Particulate Performance." Applied Materials, Inc., 1986.

Frosch, Carl J., and Lincoln Derick, "Surface Protection and Selective Masking during Diffusion in Silicon." *Journal of the Electrochemical Society,* September 1957.

Geraghty, John M., and Technology Group, "The Semiconductor Industry, A Glossary of Technical Terms." New York: Dean Witter Reynolds, Inc., 1984.

"How Intel Manufactures Integrated Circuits," Pamphlet No. 231800-002. Santa Clara, Calif.: Intel Corporation, no date.

Mann, David W., "Continuous Process Photomask System."

Burlington, Vt.: David W. Mann Company, Division of GCA Corporation, 1966.

Mills, Thomas G., "Gallium Arsenide Technology." *TRW Quest,* spring 1980.

National Materials Advisory Board, *State of the Art Reviews: Advanced Processing of Electronic Materials in the United States and Japan.* Washington: National Academy Press, 1986.

Resor, Griffith L., "Microlithography: Review of Important Trends and Equipment Alternatives." GCA Corporation, April 1985.

"The Role of Advanced CVD Technology in Semiconductor Manufacturing." Applied Materials, Inc., 1987.

"The Role of Epitaxial Deposition in Semiconductor Manufacturing." Applied Materials, Inc., 1986.

Roussel, Jeanne M., "Step-and-Repeat Comes of Age: History of Step-and-Repeat." GCA Corporation, 1982.

Schwettmann, Frederic N., and John L. Moll, "IC Process Technology: VLSI and Beyond." *Hewlett-Packard Journal,* August 1982.

"Seed to Semiconductor: An Overview." Pamphlet. Motorola Inc., Semiconductor Products Sector, 1986.

SMT-CAD Technology Corporation and L.G.M. Consulting, Inc., *Surface Mount Technology.* Cupertino, Calif.: Electronic Trend Publications, 1986.

Stidham, John, Jack Smith, and Gary Dickerson, "Automated Defect, Dimension and Registration Control in an Advanced Wafer Fab." Abstract, no date.

Strain, Robert J., "Solving Problems in Metal-Oxide Semiconductors." *Bell Laboratories Record,* October 1970.

"Texas Instruments Inventor Inducted into National Inventors' Hall of Fame." News Release No. C-492. Dallas: Texas Instruments, Inc. (Texas Instruments Archives), February 10, 1982.

Texas Instruments Learning Center, *Understanding Solid-State Electronics.* Dallas: Texas Instruments, Inc., 1978.

Trends in IC Packaging and Assembly. San Jose, Calif.: Strategic, Inc., April 1984.

Van Zant, Peter:
"Keep It Clean." Videotape. Redwood City, Calif.: Semiconductor Services, 1986.
"At the Limits." Videotape. Redwood City, Calif.: Semiconductor Services, 1986.

Acknowledgments

The editors would like to express their appreciation to the following individuals and institutions for their assistance with the preparation of this volume: **In France:** Paris—Maria Grazia Prestini, SGS-Thomson Microelectronics. **In The United States:** Arizona—Phoenix: Chuck Burnside, Burnside & Associates; California—Palo Alto: Nathalie F. Gross, Varian Associates, Inc.; Redondo Beach: VLSI Facility, TRW Electronics & Technology Division; San Jose: Gene Norrett and Penny Sur, Dataquest; Jerry D. Hutcheson, VLSI Research, Inc.; Santa Clara: Jacque Jarve, Intel Corporation; Sunnyvale: Peter Stoll, Intel Corporation; Colorado—Colorado Springs: David Smith, Inmos Corporation; Connecticut—Wilton: Keith Y. Mortensen, The Perkin-Elmer Corporation; Maryland—Gaithersburg: George G. Harman, National Bureau of Standards; Anne Ledger, Electro Mechanical Design Services, Inc.; Kensington: Jane Gruenebaum; Lutherville: Charles Harper, Technology Seminars, Inc.; Massachusetts—Andover: Jeanne Roussel and Edward J. Sweeney, GCA Corporation; Attleboro: Michael Kirkman, Augat Interconnection Systems; Norwood: William Schweber, Analog Devices; Michigan—Ann Arbor: Lynn Conway, University of Michigan; Missouri—St. Louis: Mary L. Nowicki, Hill and Knowlton, Inc.; Donald C. Otto, Monsanto Electronic Materials Company; New Hampshire—Nashua: Carl Chartier and Robert Couto, Ferrofluidics Corporation; New Jersey—Murray Hill: Thaddeus Kowalski, AT&T Bell Laboratories; Short Hills: Kevin Compton, AT&T Bell Laboratories; New York—Armonk: Paul Bergevin, IBM Corporation; Rochester: Daniel R. Ouweleen and William H. Wood, Kayex Corporation; Pennsylvania—Malvern: Albert J. Charpentier, Ensoniq, Inc.; Willow Grove: John Condit and Leon Oboler, Kulicke and Soffa Industries, Inc.; Texas—Austin: D. Marshall Andrews and William D. Stotesbery, Microelectronics and Computer Technology Corporation; Chet Freda and Dean Mosely, Motorola; Dennis Mick, 3M; Dallas: Mary Ann Toperzer, Texas Instruments; Vermont—Essex Junction: Robert C. McMahan, Sandy Smith, and George A. Sporzynski, IBM Corporation; Virginia—Springfield: Erol's, Incorporated.

Picture Credits

Credits for illustrations in this book are listed below. Credits from left to right are separated by semicolons, from top to bottom by dashes.

Cover: Art by Mark Robinson. 7: Art by Stephen R. Wagner. 8-11: Art by Stephen R. Wagner, courtesy IBM Corporation. 12-15: Art by Stephen R. Wagner. 16: Art by Mark Robinson. 20: AT&T Bell Laboratories. 21: Courtesy Texas Instruments. 24: Courtesy National Semiconductor; Wayne Miller/Magnum Photos. 25: Courtesy National Semiconductor. 27: © David Scharf/Peter Arnold, Inc., 1985. 31: Peter Yates. 32: Jon Brenneis. 35: Art by John Drummond/Aldus Freehand® Software. 36: Courtesy INMOS, Bristol. 37, 38: Art by Al Kettler. 39: Art by John Drummond/Aldus Freehand® Software. 40: Art by Al Kettler; art by John Drummond/Aldus Freehand® Software. 41: Art by John Drummond/Aldus Freehand® Software. 42: Art by Al Kettler. 43: Art by John Drummond/Aldus Freehand® Software. 44, 45: Art by John Drummond/Aldus Freehand® Software. 46, 47: Art by John Drummond/Aldus Freehand® Software, except top left, art by Al Kettler. 48, 49: Art by John Drummond/Aldus Freehand® Software. 50: Art by Mark Robinson. 53: Courtesy Intel Corporation. 58, 59: Larry Sherer, courtesy IBM Corporation; art by Douglas R. Chezem; Jack Savage, courtesy Monsanto Electronic Materials Company. 60: Art by Douglas R. Chezem. 61: Larry Sherer, courtesy Monsanto Electronic Materials Company. 64: Larry Sherer, courtesy Intel Corporation. 68: Courtesy Intel Corporation. 69: Philip Harrington/Peter Arnold, courtesy Motorola, Inc. 75-87: Art by Stephen Bauer of Bill Burrows and Associates. 88: Art by Mark Robinson. 92: IBM Research Division. 94-97: Art by Al Pagan, grids by John Drummond. 100, 101: Courtesy Racal-Redac Systems Limited, Tewkesbury, Gloucestershire. 104, 105: Art by Thomas Miller. 106: Dr. Harry Sewell, courtesy Perkin-Elmer Corporation (2); Prof. Henry I. Smith/M.I.T. 109: Art by Sam Ward. 110: Art by Sam Ward, courtesy Kulicke & Soffa Industries—art by Sam Ward, courtesy IBM Corporation—art by Sam Ward, courtesy Microelectronics and Computer Technology Corporation. 111: Art by Sam Ward—art by Al Pagan—art by Sam Ward. 112: Art by Sam Ward, courtesy Motorola, Inc.; art by Sam Ward (2). 113: Art by Sam Ward—art by Sam Ward, courtesy Motorola, Inc. 114-117: Art by Sam Ward. 118, 119: Art by Sam Ward, courtesy Analog Devices.

Index

Numerals in italics indicate an illustration of the subject mentioned.

A

Adder circuit: design of, 35, *36-49*
Advanced Micro Devices (AMD): 54-55, 73, 74
Algorithms to Silicon: 102-103
Aligners: 63
Aluminum conductors: *14-15;* on gigachip, 89-90; resistance, 27
Anderson, Philip: 91
Anderson, Richard W.: 67, 70, 73
Anelco: 53
Application-specific integrated circuits (ASICs): 101-102
Arithmetic Logic Units (ALUs): 102
Armstrong, John: 92, 93
Atalla, John: 30
AT&T: 67

B

Bardeen, John: 17, *20*
Bednorz, Johannes Georg: *92*
Bell Laboratories: 17, 20, 57, 102; and superconductivity, 91, 92, 93
Bonding: *110-111*
Bottom-Up Design (BUD): 103
Brattain, Walter: 17, *20*
Brookhaven National Laboratory: 107
Burger, Robert: 73

C

CAD (computer-aided design): for integrated circuits, 33-34, 35, 99, *100-101,* 102-104; for wiring paths within and between chips, *94-97*
Caltech-Xerox project: 31-32
Capacitor: 23
Casings: *112-113*
Charpentier, Al: 99, 101, 102
Chemical vapor deposition: 80-*81*
Chemicals: hazardous, 78
Chipmaking industry: and change, 56-57; chip yields, 55; customers, 54; geographic concentrations, 54; and government, 74; labor issues, 56; marketing by, 54-55; in 1980s, 73-74; relationships among firms, 52-54; in Silicon Valley, 52-54; survival problems, 51-52, 73
Chips. *See* Integrated circuits
Chu, Paul: 93
Circuit boards: of PC, *8-13;* structure, 11, *114-117*
Circuit design: adder, 35, *36-49;* arrays in PC, *8-9;* criteria, 28; computer-aided, 33-34, *100-101,* 102-104; difficulties

with LSI chips, 31; of high performance chips, 102-104; logic gates, 38, *39-47;* mask making, *48-49;* networks of gates, *40-41, 46-47;* planning, *37;* rules, 31-32; table of combinations, *38;* training of designers, 31, 32
CMOS circuitry: 30-31; for logic gates, *42-45*
Conway, Lynn: *31-32*
Cray, Seymour: 99
Crockett, Bruce: 99

D

De Forest, Lee: 17
Deming, W. Edwards: 67, 70-73
Derick, Lincoln: 57
Design of integrated circuits. *See* Circuit design
Dicing: *87*
Diode: 22
Doctor, Louis: 101
Doping: 19-20, 23, 80; dopants: 19-20, 57, 85; gallium arsenide, 99; process, 62, *84-85*
Dummer, G. W. A.: 19, 20-21, 22, 23

E

Electron-beam lithography: 107-108
Electronic keyboards: 100-101
Elliott, David: 107
Ensoniq, Inc.: 100-101
EPROM: interior, *12-15;* in PC, *8-9, 10-11*
Etching silicon: 57, 62, *83*

F

Fabrication: building structure, *76-77. See also* Manufacturing of chips
Fairchild Semiconductor Corporation: 18, 23, 26, 53-54, 74
Ferguson, Charles H.: 107
Frosch, Carl: 57

G

Gallium arsenide: 93; doping, 99; fabrication, 98-99
Gates, logic: 38, *39-47;* AND, *39, 44-45;* arrays in PC, *8-9;* networks, *36-37, 40-41, 46-47;* NOT, 30, *39, 43;* OR, *39, 44;* transistors in, *42-45*
GCA: 64-65
Germanium: 19, 23, 57
Gigachip: 89-90, 99, 104-108; and superconductors, 90
GTE Laboratories: 101

H

Heat: generation, 27, 28, 30; and MOSFETs, 30; removal, 28; and

superconductors, 90
Helium, liquid: 90-91
Hewlett-Packard: 70
Hoerni, Jean: 23, *24*
Hybrids: *119*

I

IBM: 54, 67; and superconductivity, 91, 92, 93; and synchrotron chipmaking, 107
Inductor: 23
Integrated circuits: *64;* advantages over wired circuitry, 24; ASICs, 101-102; circuit design, 31-34, *100-101,* 102-104; on computer motherboard, 7, *8-15;* cost of, 18; density of components, 18, 19, 30; early proposal for, 19, 21; execution time of, 18; feature size comparison, *104-105;* first, *21,* 22-23; future, 108; gallium arsenide, 93, 98; gigachip, 89, 102, 104, 108; layers, *80-81;* interconnecting, *94-97;* manufacturing difficulties, 25; and MOS technology, 30; power consumption, 31; price, 25, 26; quality control, 65-67, 70; significance, 7; structure, *8-15;* and superconductors, 90-93; surface, *27;* and transistors, 18; wires linking circuit to pins, *12-13. See also* Chipmaking Industry; Manufacturing; Transistors
Intel: 51, 53; 4004, *68*
Intersil: 53
Introduction to VLSI Systems (Mead and Conway): *31,* 32
Inverter: 30, *39, 43*
Ion implantation: 62, *84-85*

J

Japan: 67, 70, 71-72
Josephson, Brian: 91
Josephson junction: 91-92

K

Kenzler, Gene: 91
Khang, Dawon: 30
Kilby, Jack St. Clair: *21,* 22, 25
KLA Instruments: 66
Kowalski, Thaddeus: 102, 104

L

Last, Jay: *53*
Layering process: 57, *80-81*
LSI (large-scale integration): 18; design of circuits, 31-32

M

Manufacturing of chips: 75, *76-87;* advantages of small size, 26, 55; basic process, 57, *58-61, 62;* chemical handling, 78, *79;* creating masks, *48-49, 63;* difficulties, 25, 55; doping, 62, 80, *84-85;* and dust, 26, *78-79;* electron-beam lithography, 107-108; etching, 57, 62, *83;* fabrication building structure, *76-77;* gigachip technology, 104, 106-108; NMOS and PMOS transistors, 29; photolithography, 62-65, *82-83,* 104, *106;* quality control, 65-67, 70-73, *86-87;* silicon wafers, *58-61;* testing, *86-87;* vibration isolation, *76-77;* x-ray lithography, *106,* 107. *See also* Packaging

Marketing of chips: 54, 57

Masks: *48-49; 63; 65*

Mead, Carver: 31, *32*

Meindl, James D.: 104, 108

Memory chips: improvements in density, 27; interior of EPROM, *14-15;* mounting of, *10-11;* in PC, *8-9;* surface, *27*

Microchips: *See* Integrated Circuits

Micromodule Plan: 22

Microprocessor: *68-69;* first, *68;* in PC, *8-9*

Military electronics: as early market for chips, 27, 53; and soldered circuitry, 22

Miniaturization of circuits: 18, 26, 55, 104, 106-108, *118-119*

Modules: *8-9;* mounting of, *10-11*

Monolithic Memories: 56, 74

Moore, Gordon: 18, *53,* 66; and Moore's Law, 18

MOS (metal-oxide semiconductor): 29-30; types and varieties of, 29-30, 31

Motherboard, of PC: *7, 8-15*

Motorola: *26; 68030, 68, 69*

Müller, Karl Alex: *92*

N

National Semiconductor: 74

NASA: as early market for chips, 27

Nishibori, E. E.: 72

Nitrogen, liquid: 91

Noyce, Robert: 24, *25, 53, 73*

O

Onnes, Heike Kamerlingh: 90

P

PABX: 101

Packaging: 109; bonding, *110-111;* casings, *112-113;* lead arrangements, *112-113*

Patterning process: 57, *82-83*

Photolithography: 62-65, 75, *82-83,* 104, *106;* photoresist, *80, 82-83*

Project Tinkertoy: 22

Q

Quality control: 65-67; automated, 66; in Japan, 70; statistical process control, 67, 70-71; testing, *86-87;* visual inspection, 66

R

Raster Technologies: 101

Raytheon: 63

Resistor: 23

Roberts, Sheldon: *53*

Roussel, Jeanne: 65

S

Samplers: 100-101

Sanders, Jerry: 54, 73

Sandia National Laboratories: 93

Schottky, Walter: 98

Sematech: 74

Semiconductors: 19-20, 93. *See also* Gallium arsenide; Germanium; Silicon

Shewhart, Walter: 71, 72

Shockley, William: 17, 19, *20,* 29, 52, *53,* 98

Shockley Eight: 53

Shockley Semiconductor Laboratories: 52-53

Signetics: 53

Silicon: 19; as choice for semiconductor products, 23, 57; doping, 19-20, 23, *84-85;* etching, 57, 62, *83;* and gallium arsenide, 93, 98; insulating, 57; polysilicon, 42, *80-83;* wafer manufacture, *58-61*

Silicon compilers: 33, 102-104

Silicon Desert: 54

Silicon dioxide: 22; etching, 57, 62; formation of layer, 57, *80-81;* impurities in, and MOSFETs, 29-30

Silicon Mountain: 54

Silicon Prairie: 54

Silicon Valley: 52-54

Space research: and development of microchip, 25-26; and soldered circuitry, 21-22

Speed: 89-93, 97-108; and density, 89; of electric current, 90; and gallium arsenide, 93, 99; superconductivity, 90; and wiring paths, 94

Sperry Univac: 91, 92

Sputtering: 80

Statistical process control: 67, 70-72

Stepper (step-and-repeat camera): 62-63; wafer, 64-65

Sunnyvale, California: 52

Superconductivity: 90-91; Josephson junction, 91-92; magnets, 91; temperatures, 90, 91, 92-93

Synchrotrons: 107

T

Telecommunications Research Establishment, England: 19

Texas Instruments: 22, 23, 24, 26, 54

Theft of chips: 56

Tilton, John: 51

Timing: of signals on chip, 34

Transistors: bipolar, 20, 28; CMOS circuits, 30-31, *42-43;* gallium arsenide, 98; invention of, and computer, 17-18, 19; junction, 19, 20, 23-24; MOS, 29-30; MOSFETs, 29; NMOS, 29; n-p-n and p-n-p, 28; planar, 23-24; PMOS, 29; power usage, 28, 30; silicon and germanium, 19, 57, 62; structure of, 20, 29; unipolar, 29

U

Union Carbide Electronics: 53

V

Vacuum tubes: 17

Vias: *10-11, 116-117*

VLSI (very-large-scale integration): 18; design of circuits, 31-32

W

Wafers: *64;* manufacturing, *58-61;* effect of size, 63; quality control of processed, 65-66; stepper, 64-65

Wiring paths: *94-97*

X

X-ray lithography: *106,* 107

Y

Yannes, Bob, 99-100

Time-Life Books Inc.
is a wholly owned subsidiary of
THE TIME INC. BOOK COMPANY

President and Chief Executive Officer: Kelso F. Sutton
President, Time Inc. Books Direct:
Christopher T. Linen

TIME-LIFE BOOKS INC.

EDITOR: George Constable
Executive Editor: Ellen Phillips
Director of Design: Louis Klein
Director of Editorial Resources: Phyllis K. Wise
Director of Photography and Research:
John Conrad Weiser

PRESIDENT: John M. Fahey, Jr.
Senior Vice Presidents: Robert M. DeSena,
Paul R. Stewart, Curtis G. Viebranz, Joseph J. Ward
Vice Presidents: Stephen L. Bair, Bonita L.
Boezeman, Mary P. Donohoe, Stephen L. Goldstein,
Juanita T. James, Andrew P. Kaplan, Trevor Lunn,
Susan J. Maruyama, Robert H. Smith
New Product Development: Yuri Okuda,
Donia Ann Steele
Supervisor of Quality Control: James King

PUBLISHER: Joseph J. Ward

Editorial Operations
Copy Chief: Diane Ullius
Production: Celia Beattie
Library: Louise D. Forstall

Computer Composition: Gordon E. Buck (Manager),
Deborah G. Tait, Monika D. Thayer,
Janet Barnes Syring, Lillian Daniels

Correspondents: Elisabeth Kraemer-Singh (Bonn);
Christina Lieberman (New York); Maria Vincenza
Aloisi (Paris); Ann Natanson (Rome). Valuable
assistance was also provided by: Christine Hinze
(London); Elizabeth Brown (New York).

Library of Congress Cataloging in Publication Data

The Chipmakers / by the editors of Time-Life Books.—Rev. ed.
    p.   cm.—(Understanding computers)
    Includes bibliographical references (p.
    1. Integrated circuits—Very large scale integration.
    2. Microelectronics.
    I. Time-Life Books.   II. Series.
TK7874.C543 1990   621.39'5—dc20   90-10706 CIP
ISBN 0-8094-7618-5
ISBN 0-8094-7619-3 (lib. bdg.)

For information on and a full description of any Time-Life
Books series, please call 1-800-621-7026 or write:
Reader Information
Time-Life Customer Service
P.O. Box C-32068
Richmond, Virginia 23261-2068

Time-Life Books Inc. offers a wide range of fine recordings,
including a *Rock 'n' Roll Era* series. For subscription
information, call 1-800-621-7026 or write Time-Life Music,
P.O. Box C-32068, Richmond, Virginia 23261-2068.

UNDERSTANDING COMPUTERS

SERIES DIRECTOR: Lee Hassig
Series Administrator: Loretta Britten

Editorial Staff for *The Chipmakers*
Designer: Christopher M. Register
Associate Editors: Jean Crawford (pictures),
John R. Sullivan (text)
Researchers: Stephanie A. Lewis, Gwen C. Mullen,
Barbara Swanke
Writer: Robert M. S. Somerville
Assistant Designer: Tessa Tilden-Smith
Copy Coordinator: Elizabeth Graham
Picture Coordinator: Robert H. Wooldridge, Jr.
Editorial Assistant: Susan L. Finken

Special Contributors: Joseph Alper, Elizabeth
Carpenter, Bonnie Gordon, Richard Jenkins, Martin
Mann, Eugene Rodgers, Joel Shurkin (text); Robert M.
McDowell, Chris Russell, Julie Ann Trudeau
(research); Mel Ingber (index)

THE CONSULTANTS

DAVID ANDREWS, an IBM engineer since 1967, has
designed circuit boards for a number of IBM products,
including the Personal System II Model 30 computer.

ROBERT M. BURGER is a vice president and chief scientist of the Semiconductor Research Corporation in Research Triangle Park, North Carolina.

EDWARD M. DAVIS serves as president of the Virginia
Center for Innovative Technology and is a former president of IBM's General Technology Division.

DAVID J. ELLIOTT, vice president for marketing and sales
of Leitz-Image Micro Systems Co., Billerica, Massachusetts, is the author of several books on integrated-circuit
technology.

BRIAN HAYES has written about computers and computing for *Scientific American, BYTE,* and *Computer Language.* He is author of a book about Scheme, a dialect of
the LISP programming language.

STAN KIRSCHNER is staff assistant to the plant manager
at IBM's integrated circuit fabrication plant in Essex Junction, Vermont.

FREDERICK B. MAXWELL, a microcontroller specialist, is
presently working under contract to the United States
Postal Service in its Process Control Division in Landover,
Maryland.

JAMES D. MEINDL is senior vice president for academic
affairs and provost of Rensselaer Polytechnic Institute.
Formerly, he directed Stanford University's Center for
Integrated Systems.

ARNOLD PFHANL has doctorates in physics from Graz
University in Austria and from the University of Paris.
Holder of several French and American patents, he has
been with AT&T Bell Laboratories since 1957.

ROBERT I. SCACE is deputy director of the National
Bureau of Standards Center for Electronics and Electrical
Engineering. He has worked in the semiconductor field
since 1954.

KENNETH A. TASCHNER, an engineer with Pacific Scientific, an electronics firm in Silver Spring, Maryland, sits
on the advisory board of the Capitol Institute of Technology in Laurel, Maryland.

PETER VAN ZANT is a California-based training and
communications consultant to the semiconductor industry.

REVISIONS STAFF

EDITOR: Lee Hassig

Writer: Esther Ferington
Assistant Designer: Tina Taylor
Copy Coordinator: Anne Farr
Picture Coordinator: Katherine Griffin
Consultant: Robert M. Burger